Machine Learning with PySpark

With Natural Language Processing and Recommender Systems

Second Edition

Pramod Singh

Apress®

Machine Learning with PySpark

Pramod Singh
Bangalore, Karnataka, India

ISBN-13 (pbk): 978-1-4842-7776-8   ISBN-13 (electronic): 978-1-4842-7777-5
https://doi.org/10.1007/978-1-4842-7777-5

Managing Director, Apress Media LLC: Welmoed Spahr
Acquisitions Editor: Celestin Suresh John
Development Editor: Mark Powers
Coordinating Editor: Aditee Mirashi

Cover designed by eStudioCalamar

Cover image designed by Freepik

Distributed to the book trade worldwide by Springer Science+Business Media New York, 1 New York Plaza, Suite 4600, New York, NY 10004-1562, USA. Phone 1-800-SPRINGER, fax (201) 348-4505, e-mail orders-ny@springer-sbm.com, or visit www.springeronline.com. Apress Media, LLC is a California LLC and the sole member (owner) is Springer Science + Business Media Finance Inc (SSBM Finance Inc). SSBM Finance Inc is a **Delaware** corporation.

For information on translations, please e-mail booktranslations@springernature.com; for reprint, paperback, or audio rights, please e-mail bookpermissions@springernature.com.

Apress titles may be purchased in bulk for academic, corporate, or promotional use. eBook versions and licenses are also available for most titles. For more information, reference our Print and eBook Bulk Sales web page at http://www.apress.com/bulk-sales.

Any source code or other supplementary material referenced by the author in this book is available to readers on GitHub via the book's product page, located at www.apress.com/978-1-4842-7776-8. For more detailed information, please visit http://www.apress.com/source-code.

Printed on acid-free paper

I dedicate this book to my wife Neha, my son Ziaan, and my parents. Without you guys, this book wouldn't have been possible. You complete my world and are the source of my strength.

Table of Contents

About the Author

Pramod Singh works at Bain & Company as a Senior Manager, Data Science, in their Advanced Analytics group. He has over 13 years of industry experience in Machine Learning (ML) and AI at scale, data analytics, and application development. He has authored four other books including a book on Machine Learning operations. He is also a regular speaker at major international conferences such as Databricks AI, O'Reilly's Strata, and other similar conferences. He is an MBA holder from Symbiosis International University and data analytics certified professional from IIM Calcutta. He lives in Gurgaon with his wife and five-year-old son. In his spare time, he enjoys playing guitar, coding, reading, and watching football.

About the Technical Reviewer

Joos Korstanje is a data scientist, with over five years of industry experience in developing machine learning tools, of which a large part is forecasting models. He currently works at Disneyland Paris where he develops machine learning for a variety of tools.

Acknowledgments

I am really grateful to Apress for providing me with the opportunity to author the second edition of this book. It's a testament that readers found the first edition really useful. This book wouldn't have seen the light of day if a few people were not there with me during this journey. First and foremost, I want to thank the most important person in my life, my beloved wife, Neha, who selflessly supported me throughout this time and sacrificed so much just to ensure that I complete this book.

I want to thank Dominik Utama (partner at Bain) for taking out time from his busy schedule to write the foreword for the readers of this book. Thanks, Dom. I really appreciate it. I would also like to thank Suresh John Celestin who believed in me and offered me to write the second edition. Aditee Mirashi is one of the best editors to work with – this is my fourth book with her. She is extremely supportive and responsive. I would like to especially thank Mark Powers, who dedicated his time for reading every single chapter and giving so many useful suggestions. Another person I want to thank is Joos Korstanje who had the patience to review every single line of code and check the appropriateness of each example. Thank you, Joos, for your feedback and your encouragement. It really made a difference to me and the book as well.

I also want to thank my mentors who have constantly guided me in the right direction. Thank you, Dominik Utama, Janani Sriram, Barron Berenjan, Ilker Carikcioglu, Dr. Vijay Agneeswaran, Sreenivas Venkatraman, Shoaib Ahmed, and Abhishek Kumar, for your time.

Finally, I am grateful to my son, Ziaan, and my parents for the endless love and support they give me irrespective of circumstances. You guys remind me that life is beautiful.

Foreword

Businesses are swimming in data, yet many organizations are struggling to remain afloat in a rapidly expanding sea of data. Ever-increasing connectivity of data-transmitting devices is driving growth in how much and how fast data is being generated. This explosion in digitalization has been accompanied by a proliferation of applications and vendors. Companies often use multiple vendors for the same use cases and store data across multiple systems. Digital data lives in more and more formats and across fragmented data architecture layers.

In a world where data lives everywhere, many organizations are on a race to be more data-driven in their decision-making to get and stay ahead of the competition. The winners in this race proactively manage their data needs and leverage their data and analytics capabilities to drive tangible business outcomes.

The starting point for using data and analytics as a strategic tool is having good data. Take a B2B company that was looking to improve how it forecasted next month's sales, for example. Each month this company would aggregate individual sales reps' "hot leads" into a sales forecast that might be wide of the mark. Improving the underlying data quality required four changes to how the company captured customer insights from the front line. First, the company clearly defined at which stage in the sales funnel a lead should be tagged to the sales forecast. Second, it reduced the granularity of information that the sales force had to enter along with each sales lead. Third, it gave more time to the sales reps to complete the data entry. Fourth, it secured commitment from the sales force to improve the data quality. Along with ensuring a common understanding of the changes, the company was able to improve how it translated frontline customer insights into better data.

With better data in their arsenals, businesses need to orchestrate the data in a complex web of systems, processes, and tools. Despite having a state-of-the art CRM (customer relationship management) system, numerous spreadsheets were metronomically circulated across the organization to syndicate the company's sales forecast. The company improved how it orchestrated the sales forecast data in four ways. First, it consolidated disparate planning and tracking tools into a consistent view in the CRM system. Second, it secured commitment from the organization to use the CRM forecast as one source of truth. Third, it configured the access for all relevant stakeholders to view the sales forecast.

Fourth, it compressed the organizational layers that could make adjustments – as fewer layers brought fewer agendas in managing the sales forecast.

The effective use of data and analytics can help businesses make better decisions to drive successful outcomes. For instance, the company mentioned previously developed a system of sanity checks to assure the sales forecast. It used machine learning to classify individual sales reps into groups of optimistic and conservative forecasters. Machine learning helped predict expected conversion rates at an individual level and at an aggregate basis that the company could use to sense-check deviations from the forecast envelope, either to confirm adjustments to the sales forecast or trigger a review of specific forecast data. Better visibility on the sales forecast enabled the supply chain function to proactively move the company's products to the branches, where sales was forecasted to increase, and place replenishment orders with its suppliers. As a result, the company incurred fewer lost sales episodes due to stock-outs and achieved a more optimal balance between product availability and working capital requirements.

Building data and analytics capabilities is a journey that comes in different waves. Going back to the example, the company first concentrated on a few use cases – such as improved sales forecasting – that could result in better visibility, more valuable insights, and automation of work. It defined and aligned on changes to its operating model and data governance. It set up a center of excellence to help accelerate use cases and capability building. It built a road map for further use case deployment and capability development – upskilling current employees, acquiring new talent, and exploring data and analytics partnerships – to get the full value of its proprietary data assets.

Great data and analytics capabilities can eventually lead to data monetization. Companies should start with use cases that address a raw customer need. The most successful use cases typically center on improving customer experience (as well as supplier and frontline worker experience). Making the plunge to build data and analytics capabilities is hard; yet successful companies are able to muster the courage to ride the waves. They start with small wins and then scale up and amplify use cases to capture value and drive tangible business outcomes through data and analytics.

Foreword by Dominik Utama

Introduction

I am going to be very honest with you. When I signed the contract to write this second edition, I thought it would be a bit easier to write, but I couldn't have been more wrong about this assumption. It has taken me quite a significant amount of time to complete the chapters. What I have come to realize is that it's never easy to break down a thought process and put it on paper in the most convincing manner. There are so many retrials in that process, but what helped was the foundation block or the blueprint that was already established in the first edition of this book. The main challenge was to figure out how I could make this book more relevant and useful for the readers. I mean there are literally thousands of books on this subject already that this might just end up as another book on the shelf.

To find the answer, I spent a lot of time thinking and going through the messages that I received from so many people who read the first edition of the book. After a while a few patterns started to emerge. The first realization was that data continues to get generated at a much faster pace. The basic premise of the first edition was that a data scientist needs to get familiar with at least one big data framework in order to handle the scalable ML engagement. It would require them to gradually move away from libraries like sklearn that have certain limitations in terms of handling large datasets. That is still highly relevant today as businesses want to leverage as much data as possible to build powerful and significant insights. Hence, people would be excited to learn new things about the Spark framework.

Most of the books that have been published on this subject were either too detailed or lacked a high-level overview. Readers would start really easy but, after a couple of chapters, would start to feel overwhelmed as the content became too technical. As a result, readers would give up without getting enough out of the book. That's why I wanted to write this book that demonstrates the different ways of using Machine Learning without getting too deep, yet capturing the complete methodology to build a ML model from scratch.

Another issue that I wanted to address in this edition is the development environment. It was evident many people struggled with setting up the right environment in their local machines to install Spark properly and could see a lot of issues. Hence, I wrote this edition using Databricks as the core development platform, which is easy to access, and one doesn't have to worry about setting up anything on the local system. The best thing about using Databricks is that it provides a platform to code in multiple languages such as Python, R, and Scala. The other extension to this edition is that the codebase demonstrates end-to-end development of ML models including automating the intermediate steps using Spark pipelines. The libraries that have been used are from the latest Spark version.

This book is divided into three different sections. The first section covers the process to access Databricks and alternate ways to use Spark. It goes into architecture details of the Spark framework, along with an introduction to Machine Learning. The second section focuses on different machine learning algorithm details and executing end-to-end pipelines for different use cases in PySpark. The algorithms are explained in simple terms for anyone to read and understand the details. The datasets that are being used in the book are relatively smaller on scale, but the overall process and steps remain the same on big data as well. The third and final section showcases how to build a distributed recommender system and Natural Language Processing in PySpark. The bonus part covers creating and visualizing sequence embeddings in PySpark. This book might also be relevant for data analysts and data engineers as it covers steps of big data processing using PySpark. The readers who want to make a transition to the data science and Machine Learning field would find this book easier to start with and can gradually take up more complicated stuff later. The case studies and examples given in the book make it really easy to follow along and understand the fundamental concepts. Moreover, there are limited books available on PySpark, and this book would certainly add value towards upskilling of the readers. The strength of this book lies in explaining the Machine Learning algorithms in most simplistic ways and taking a practical approach toward building and training them using PySpark.

I have put my entire experience and learnings into this book and feel it is precisely relevant to what readers are seeking either to upskill or to solve ML problems. I hope you have some useful takeaways from this book.

CHAPTER 1

Introduction to Spark

This is the introductory chapter to Spark in order to set the initial foundation for the rest of the chapters. This chapter is divided into three parts – understanding the evolution of data, core fundamentals of Spark along with its underlying architecture, and different ways to use Spark. We start by going through the brief history of data generation and how it has evolved in the last few decades. If we were to compare today, it certainly looks different from the days when the internet was still new and bursting upon the scene. Things have changed in so many ways over the last 20 years that it is hard to encapsulate everything in this introductory paragraph. Perhaps, one of the most prominent changes that stands out in this context is the amount of data being generated nowadays. People around the world are now using multiple applications on different digital devices that make their lives easier. These apps have opened up doors for the users to have instant access to things and hence immediate gratification – for example, ordering a chicken burger, fixing an appointment with a doctor, or registering for a yoga class the same day. For users of these apps, it is just a medium to interact with the platform and place the request, but in the background, it generates a series of data points. These data points could be a set of steps the user has taken on the app and associated information. As a result of multiple users accessing these apps, enormous data gets generated. Then come different devices and sensors that also churn out data at frequent intervals known as IoT devices. Now this data generated at the back end needs to be cleaned, aggregated, and persisted to help and make these apps more intelligent and relevant for the users. As they say, "data" is the new oil. Let's jump into the first section of this chapter.

© Pramod Singh 2022
P. Singh, *Machine Learning with PySpark*, https://doi.org/10.1007/978-1-4842-7777-5_1

Data Generation

In order to understand the evolution of data, let us divide the overall timeline into three parts:

- Before the 1990s

- The Internet and Social Media Era

- The Machine Data Era

Before the 1990s

In those days, data was generated or accumulated by data entry operators. In essence, only the employees of different organizations could enter the data into these monolithic systems. Also, the number of data points were very limited, capturing only a few fields. For example, a bank or an insurance company would collect user data such as annual income, age, gender, and address using some physical forms and later provide it to the data entry person who would insert the details into the customer database (DB). Till this stage, the data was limited, and hence legacy databases were enough to run the show. The companies were sitting on this data, which wasn't used much beyond the required cases. There were very basic applications that were running at that time that used the customer data. Most of these were limited to only internal use. Data was primarily stored for reference and validation purposes only.

The Internet and Social Media Era

Then came the phase where the Internet got widely adopted and information was made easily accessible to everyone who used the Internet. In comparison with the early stage, the Internet users could now provide and generate their own data. This was a tremendous shift as the number of Internet users grew exponentially over the years, and hence the data generated by the users increased at an even higher rate as shown in Figure 1-1.

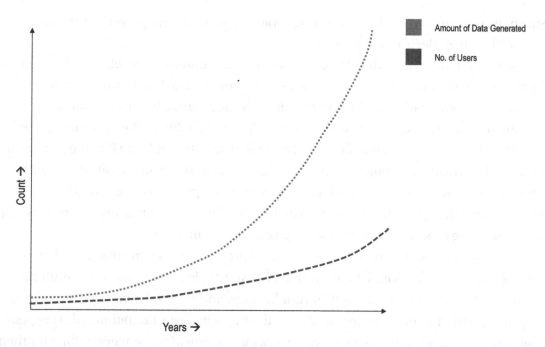

Figure 1-1. *Data growth in the Internet Era*

For example, the users could now log in/sign up over the Internet by providing their own details and upload photos and videos on various social platforms. The users could go on different platforms to leave their comments and reviews for different products and services. In essence, multiple data points started getting generated from each user, and companies started to store the user data through websites or apps that included user decision, usage patterns, feedback, etc. Now this data was a mix of structured and unstructured forms, and the companies simply couldn't rely on traditional databases and hence also migrated to NoSQL DBs. The size of this new data was already so huge that a group of machines (known as clusters) were required to process it.

The Machine Data Era

We are currently in the Machine Data Era where machines/devices around us are generating data at an extremely fast rate. Things like smartphones, tablets, laptops, cars, and heavy machinery using sensors are populating data at every microsecond, which gets captured for different purposes. This data is even higher in magnitude than the user-generated data. In today's economy, "data" is supercritical for companies to have the competitive edge over their competitors. Hence, a lot of third-party data is procured externally to augment the data crunching capabilities and derive powerful insights to take key business decisions and so on.

As mentioned earlier, when the data was still at the enterprise level, the relational database was good enough to handle the needs of the system, but as the size of data increased exponentially over the past couple of decades, there has been a pulsating need of looking at the approach to handle the big data in a different way, which has led to the birth of Spark. Traditionally, we used to take the data and bring it to the processor because data wasn't very huge in size and volume, but now it would easily overwhelm the processor. Hence, instead of bringing the data to the processor, we reverse the approach and bring multiple processors to the data. This is known as parallel processing as data is being processed at a number of places at the same time.

Let's take an example to understand parallel processing. Assume that you plan to travel and reach the airport to board your flight. Now typically you would see multiple counters for you to be able to check in your luggage, and it would take just a few minutes for you to finish the security check and go to the boarding gate. Sometimes, during peak season or holiday season, one might observe a long queue of passengers waiting for their turn on the counters. There are two reasons: a higher number of passengers flying during peak season or some of the counters being not operational. We can think of numerous such examples – waiting time at McDonald's if just a single kiosk is operational vs. if multiple kiosks are running.

Clearly, if things can be run in parallel, they are much faster to execute. Parallel processing works on a similar principle, as it parallelizes the tasks and accumulates the final results at the end. Spark is a framework to handle massive datasets with parallel processing at high speed and with a robust mechanism. As a result, it is much faster and scalable compared with other traditional approaches. Now, there are certain overheads that come along with using Spark, but we will get into some of the limitations of Spark in the later chapters. Let's go over Spark's evolution and its architecture in detail in the next section.

Spark

Moving on to the history of Apache Spark. If we refer to the evolution timeline as shown in Figure 1-2, we will notice that it started as a research project at the UC Berkeley AMPLab in 2009 and was open sourced in early 2010. It was then donated to the Apache Software Foundation in 2013. The next few years were very important for Apache Spark as APIs for Dataframe and Machine Learning were introduced and got widely adopted very quickly. In 2016, Spark released TensorFrame to provide the route for Deep Learning models for data scientists/ML engineers. In 2018 we witnessed version 2.3 that introduced Kubernetes support – which meant that Spark now could be run on a Kubernetes Cluster as well.

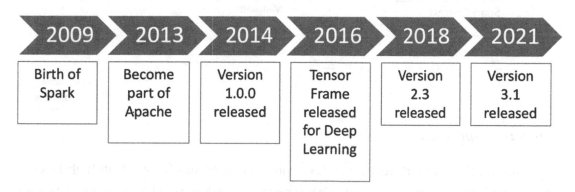

Figure 1-2. *Spark evolution*

Today, you would see Spark being used at many entities and different industries that are building numerous apps/products using Spark under the hood. By default, Apache Spark is considered to be the top choice for tackling various big data challenges. With more than 500 contributors and a user base of 0.3M+ members, Apache Spark has become a mainstream framework across all key industries. This year, they have come up with version 3.1 that brings a whole lot of improvements in terms of ML offerings, structured streaming, and other compatibilities.

The prime reason that Spark is hugely popular is the fact that Spark is very easy to use for data processing, Machine Learning, and streaming data and it's comparatively very fast since it does all in-memory computations. Since Spark is a generic data processing engine, it can easily be used with various data sources such as HBase, Cassandra, Amazon S3, HDFS, etc. Spark provides the users four language options for its use: Java, Python, Scala, and R.

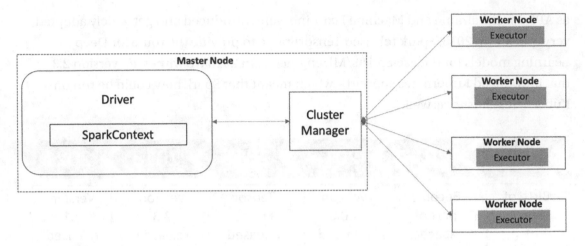

Figure 1-3. *Spark architecture*

Coming to the core architecture and key components of Apache Spark, its high-level architecture is pretty straightforward. We just saw that Spark is based on the core principles of distributed and cluster computing, and hence it works similar to a master-worker operating model. The master is responsible for assigning the tasks, and workers execute those tasks in parallel. So on a high level, Spark's architecture contains three main blocks as shown in Figure 1-3:

1. The Master process

2. The Worker Process

3. The Cluster Manager process

The Master process consists of something known as a driver program, which is the entry point to Spark. This driver program is responsible to run the main() function of the application. As a result, SparkContext is created. Spark Driver contains critical components such as DAG Scheduler and Task Scheduler to translate the code into jobs to be executed by the workers. Next is the Worker process, which contains multiple executors that are responsible for the execution of tasks. The code to execute any of the activity is first written on Spark Driver and is shared across worker nodes where the data actually resides.

Executors perform all the data processing–related tasks in parallel. The main role of executors is to read and write data to different sources after the in-memory computation and aggregation of final results. Finally, the Cluster Manager process helps to keep a check on the availability of various worker nodes for the next task allocation. Typically, cluster managers such as Hadoop YARN, Apache Mesos, and Spark's stand-alone cluster manager are referred to as the cluster manager service.

With respect to the core components of Spark's ecosystem, the most fundamental level contains the Spark Core, which is the general execution engine for the Spark platform. The rest of the functionalities are built on top of it as shown in Figure 1-4. The Spark Core is the backbone of Spark's supreme functionality features. It provides in-memory computing capabilities to deliver results at a great speed and supports a wide variety of applications. In terms of language support, Spark supports Python, Scala, Java, and R for development.

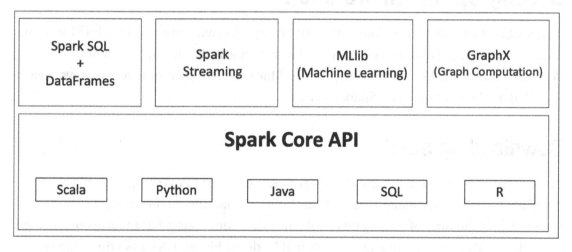

Figure 1-4. *Spark ecosystem*

Spark SQL: Many data analysts, data engineers, and scientists still prefer to work in SQL, and that's where the Spark SQL module shines through. It allows users to run interactive SQL queries for data processing. It also integrates seamlessly with the other components such as MLlib.

Spark Streaming: Sometimes things need to be done and executed in real time, for example, hyper-personalization. In such cases, batch data or historical data is not enough to build the application. One needs to apply transformations on real-time data. Spark Streaming is the module that allows the users to build real-time applications as it is capable of processing and analyzing data in near real time.

MLlib: This module is for Machine Learning at scale. Some of you who have faced issues in building large models using scikit-learn would definitely benefit from this module. It allows to build ML models on large datasets at a much higher speed, and majority of the ML algorithms are part of this module.

7

GraphX: This is a graph computation engine built on top of Spark that enables users to interactively build, transform, and reason about graph-structured data at scale. It comes with a library of common algorithms in this space.

Now that we have a good understanding of Apache Spark's core architecture and key components, we can move on to setting up our environment to be able to use Spark. Please feel free to skip the section in case you already have Spark installed on your machine.

Setting Up the Environment

This section of the chapter covers setting up the Spark environment on the local system. Spark is written in Scala, and it is packaged in such a way that it can run on both Windows and UNIX-like systems (e.g., Linux, Mac OS). The prerequisite is to have Java installed on the system to run Spark locally.

Downloading Spark

The Download section of the Apache Spark website (`http://spark.apache.org/downloads.html`) has detailed instructions for downloading the prepackaged Spark binary file. At the time of writing this book, the latest version is 3.1.1. In terms of package type, choose the one with the latest version of Hadoop. Figure 1-5 shows the various options for downloading Spark. The easiest way to get started on downloading the prepackaged binary file is clicking the link on line item #3, which will trigger the binary file download.

Download Apache Spark™

1. Choose a Spark release: 3.1.1 (Mar 02 2021) ◇

2. Choose a package type: Pre-built for Apache Hadoop 2.7 ◇

3. Download Spark: spark-3.1.1-bin-hadoop2.7.tgz

4. Verify this release using the 3.1.1 signatures, checksums and project release KEYS.

Figure 1-5. *Spark download*

Installing Spark

Once the file successfully downloaded on the local system, the next step is to unzip it. Use the following command to unzip the downloaded file:

```
[In]: tar xvf spark-3.1.1-bin-hadoop2.7.tgz
```

For Windows computers, you can use either the WinZip or 7-Zip tool to unzip the downloaded file. Once the unzip is completed, you would observe the Spark directory with the same name "spark-3.1.1-bin-hadoop2.7". Inside this Spark directory, there exist multiple folders with key files and dependencies to run Spark. The next step is to instantiate Spark shell, which gives the access to interact with and use Spark. As mentioned in the preceding section, Spark shell is in Python as well as Scala. Since most of the chapters are about PySpark, we will spin up Spark's Python shell.

To start up Spark Python shell, enter the following command in the Spark directory:

```
[In]: ./bin/pyspark
```

After a few seconds, you should see the Spark session available as shown in Figure 1-6.

```
Using Spark's default log4j profile: org/apache/spark/log4j-defaults.properties
Setting default log level to "WARN".
To adjust logging level use sc.setLogLevel(newLevel). For SparkR, use setLogLevel(newLevel).
Welcome to
      ____              __
     / __/__  ___ _____/ /__
    _\ \/ _ \/ _ `/ __/  '_/
   /__ / .__/\_,_/_/ /_/\_\   version 3.1.1
      /_/
```

Figure 1-6. *Spark Python shell*

For the purpose of learning Spark, Spark shell is a very convenient tool to use on your local computer anytime and anywhere. It doesn't have any external dependencies, other than the data files that you process will need to reside on your computer.

Docker

If you don't want to take the pain of downloading and installing Spark on your local system, then you can explore the alternative – Docker. It allows you to use Spark through containers without any need to install Spark locally. The prerequisite for this approach is to have Docker installed on your local machine. The steps to run Spark using Docker are very straightforward:

1. Ensure Docker is installed on your local system.

2. Open the terminal and run the following command:

```
[In]: docker run -p 8888:8888 jupyter/pyspark-notebook
```

3. Go to the localhost and access port 8888 to get access to the Jupyter notebook with Spark.

4. Import pyspark.

Databricks

Another way of using Spark is through Databricks' Collaborative Notebooks. Databricks is the company founded by the creators of Spark to offer commercialized and well-managed Spark-based solutions to different businesses. The Databricks platform is a powerful way to quickly perform exploratory data analysis, as well as build machine learning models along with data visualization and model management.

The Collaborative Notebooks product has two versions, the commercial and the community edition. The commercial edition is a paid product for companies to leverage all the advanced features. The community edition is free and ideal for those who would like to try out Spark and use the Databricks platform. In this part, we will cover the community edition to keep things simple. It provides an easy and intuitive environment to learn Spark, perform data analysis, or build Spark applications. You can also refer to the Databricks user guide (`https://docs.databricks.com/user-guide/index.html`) for in-depth exploration.

The first step to use Collaborative Notebooks is to sign up for a free account on the community edition at `https://databricks.com/try-databricks` as shown in Figure 1-7.

Figure 1-7. *Databricks Sign Up*

After a successful sign-up, you would land on the Databricks homepage showcased in Figure 1-8.

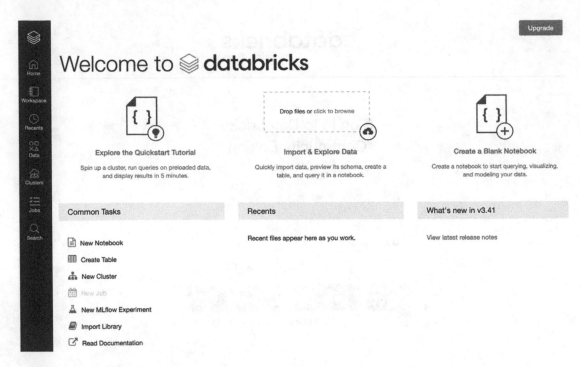

Figure 1-8. *Databricks homepage*

Spin a New Cluster

Databricks has expanded on the free cluster size from 6 GB to 15 GB for the community edition. It's actually a single node cluster that is hosted on AWS cloud. The users don't get any flexibility to tweak any cluster configuration in the free edition. A new cluster can be created under the Clusters menu as shown in Figure 1-9.

Figure 1-9. *Databricks cluster*

The next step is to click the "Create Cluster" button to spin up a single node cluster as shown in Figure 1-10. It can take anywhere between 2 minutes and 5 minutes for the cluster to become active. Once a Spark cluster is successfully initiated, a green dot appears next to the cluster name, and a Databricks notebook could now be attached to it.

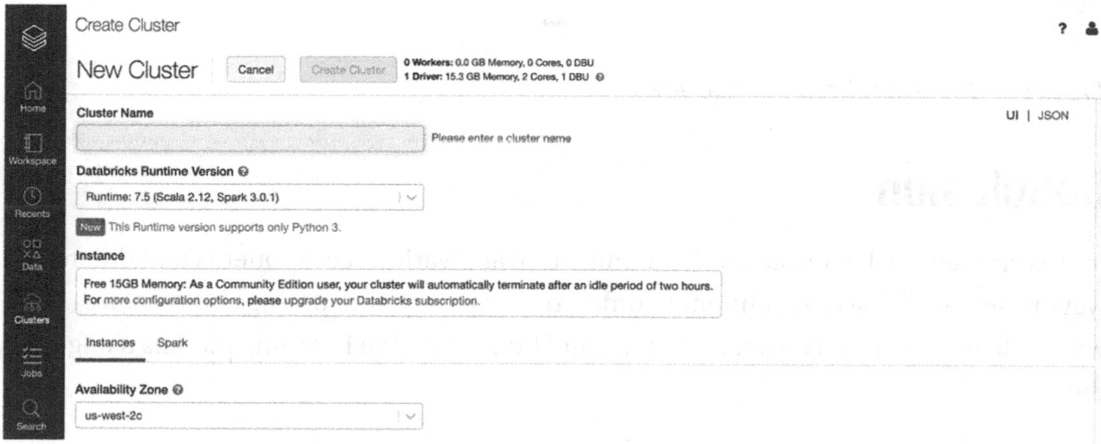

Figure 1-10. *Databricks cluster form*

13

Create a Notebook

Now that the cluster is up, we can go to the Workspace menu and create a new notebook. There are options available to import and export notebooks as well as shown in Figure 1-11. The notebook is pretty interactive and self-intuitive if you have worked with Jupyter notebooks before. The way to start writing Spark code in the notebook is to first attach the notebook to the cluster that was created previously. We can load data files of different formats such as .csv, .parquet, etc. to be used in the Databricks notebook.

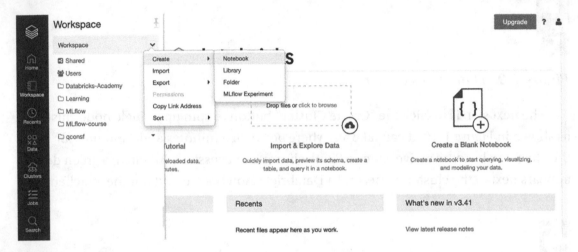

Figure 1-11. *Databricks notebook*

Conclusion

In this chapter, we looked at Spark's architecture and various components and different ways to set up a local environment in order to use Spark. In upcoming chapters, we will go deep into various aspects of Spark and build Machine Learning models using the same.

CHAPTER 2

Manage Data with PySpark

In the previous chapter, we looked at the core strength of the Spark framework and the process to use it in different ways. This chapter focuses on how we can use PySpark to handle data. In essence, we would apply the same steps when dealing with a huge set of data points; but for demonstration purposes, we will consider a relatively small sample of data. As we know, data ingestion, cleaning, and processing are supercritical steps for any type of data pipeline before data can be used for Machine Learning or visualization purposes. Hence, we would go over some of the built-in functions of Spark to help us handle big data. This chapter is divided into two parts. In the first part, we go over the steps to read, understand, and explore data using PySpark. In the second part, we explore Koalas, which is another option to handle big data. For the entire chapter, we will make use of a Databricks notebook and sample dataset.

Load and Read Data

We need to ensure that the data file is uploaded successfully in the Databricks data layer. There are multiple data file types that could be uploaded in Databricks. In our case, we upload using the "Import & Explore Data" option as shown in Figure 2-1.

© Pramod Singh 2022
P. Singh, *Machine Learning with PySpark*, https://doi.org/10.1007/978-1-4842-7777-5_2

Figure 2-1. *Upload data*

We can upload the data file "sample_data.csv" in order to read it in the Databricks notebook. The default file location would be mentioned at the bottom as shown in Figure 2-2.

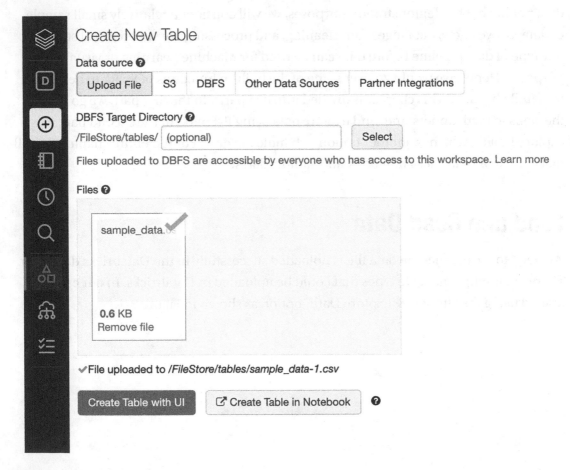

Figure 2-2. *Access data*

The next step is to spin up a cluster and attach a new Databricks notebook to it. We can go to the <u>Compute option</u> and start a new cluster. For those who might be not familiar with a cluster, it's a group of machines on a cloud platform such as AWS or Google Cloud Platform. The size of the cluster depends on the tasks and amount of data one is dealing with. There are clusters with hundreds of machines, but in our case, we would be leveraging the default cluster provided by Databricks.

Once the cluster is active, we can create a new notebook. The first set of commands are to import PySpark and instantiate SparkSession in order to use PySpark:

```
[In]: import pyspark
[In]: from pyspark.sql import SparkSession
[In]: spark=SparkSession.builder.appName("DatawithPySpark").getOrCreate()
```

Now that we have Spark up and running, we can mention the data file location along with other details:

```
[In]: file_location = "/FileStore/tables/sample_data.csv"
[In]: file_type = "csv"
[In]: infer_schema = "true"
[In]: first_row_is_header = "true"
[In]: delimiter = ","
```

We can update the read format argument in accordance with the file format (.csv, JSON, .parquet, table, text). For a tab-separated file, we need to pass an additional argument while reading the file to specify the delimiter (sep='\t'). Setting the argument inferSchema to true indicates that Spark in the background will infer the datatypes of the values in the dataset on its own:

```
[In]:df=spark.read.format(file_type).option("inferSchema",infer_schema)
.option("header", first_row_is_header).option("sep", delimiter).load
(file_location)
```

The preceding command creates a Spark dataframe with the values from our sample data file. We can consider this as an Excel spreadsheet in tabular format with columns and headers. We can now perform multiple operations on this Spark dataframe. We can use the display function to look at the dataset:

```
[In]: display(df)
```

▸ ▦ df: pyspark.sql.dataframe.DataFrame = [ratings: integer, age: integer ... 3 more fields]

	ratings ▲	age ▲	experience ▲	family ▲	mobile ▲
1	3	32	9	3	Vivo
2	3	27	13	3	Apple
3	4	22	2.5	0	Samsung
4	4	37	16.5	4	Apple
5	5	27	9	1	MI
6	4	27	9	0	Oppo

Showing all 33 rows.

We can also make use of the show function, which displays the data fields from the top row. Pandas users heavily make use of functions such as head and tail to sneak a peek into the data. Well, those are applicable here as well, but the output might look a bit different compared with display or show as it prints it in forms of rows. We can also pass the number of rows to be displayed by passing the value of n as an argument to show/head/tail:

```
[In]: df.show(5)
```

▸ (1) Spark Jobs

```
+-------+---+----------+------+-------+
|ratings|age|experience|family| mobile|
+-------+---+----------+------+-------+
|      3| 32|       9.0|     3|   Vivo|
|      3| 27|      13.0|     3|  Apple|
|      4| 22|       2.5|     0|Samsung|
|      4| 37|      16.5|     4|  Apple|
|      5| 27|       9.0|     1|     MI|
+-------+---+----------+------+-------+
only showing top 5 rows
```

```
[In]: df.head(5)

    Out[7]: [Row(ratings=3, age=32, experience=9.0, family=3, mobile='Vivo'),
      Row(ratings=3, age=27, experience=13.0, family=3, mobile='Apple'),
      Row(ratings=4, age=22, experience=2.5, family=0, mobile='Samsung'),
      Row(ratings=4, age=37, experience=16.5, family=4, mobile='Apple'),
      Row(ratings=5, age=27, experience=9.0, family=1, mobile='MI')]
```

```
[In]: df.tail(3)

    Out[8]: [Row(ratings=3, age=27, experience=6.0, family=0, mobile='MI'),
      Row(ratings=4, age=22, experience=6.0, family=1, mobile='Oppo'),
      Row(ratings=4, age=37, experience=6.0, family=0, mobile='Vivo')]
```

We can print the list of column names that are present in the Dataframe using the "columns" method. As we can see, we have five columns in the Dataframe. We can observe the datatype for each of the columns using dtypes. Finally, one of the easiest ways to understand the Dataframe is to use printSchema, which provides a list of columns along with their datatypes:

```
[In]: df.columns

      Out[9]: ['ratings', 'age', 'experience', 'family', 'mobile']
```

```
[In]: df.dtypes

                    Out[10]: [('ratings', 'int'),
                      ('age', 'int'),
                      ('experience', 'double'),
                      ('family', 'int'),
                      ('mobile', 'string')]
```

```
[In]: df.printSchema()

          root
           |-- ratings: integer (nullable = true)
           |-- age: integer (nullable = true)
           |-- experience: double (nullable = true)
           |-- family: integer (nullable = true)
           |-- mobile: string (nullable = true)
```

We can use the count method to get the total number of records in the Dataframe. We can also use describe to understand statistical measures of different numerical columns in the Dataframe. For numerical columns, it returns the measure of center and spread along with the count. For non-numerical columns, it shows the count and the min and max values, which are based on an alphabetic order of those fields and doesn't signify any real meaning. We could also pass specific column names in order to view the details of only selected data fields:

```
[In]: df.count()
```

▶ (2) Spark Jobs

```
Out[12]: 33
```

```
[In]: len(df.columns)
```

```
Out[13]: 5
```

```
[In]: df.describe().show()
```

▶ (2) Spark Jobs

```
+-------+------------------+------------------+------------------+------------------+------+
|summary|           ratings|               age|        experience|            family|mobile|
+-------+------------------+------------------+------------------+------------------+------+
|  count|                33|                33|                33|                33|    33|
|   mean|3.5757575757575757|30.484848484848484|10.303030303030303|1.8181818181818181|  null|
| stddev|1.1188806636071336|  6.18527087180309| 6.770731351213326|1.8448330794164254|  null|
|    min|                 1|                22|               2.5|                 0| Apple|
|    max|                 5|                42|              23.0|                 5|  Vivo|
+-------+------------------+------------------+------------------+------------------+------+
```

```
[In]: df.describe('age').show()
```

▶ (2) Spark Jobs

```
+-------+------------------+
|summary|               age|
+-------+------------------+
|  count|                33|
|   mean|30.484848484848484|
| stddev|  6.18527087180309|
|    min|                22|
|    max|                42|
+-------+------------------+
```

Moving on to the next data processing step – filtering data. It is a very common requirement to filter records based on conditions. This helps in cleaning the data and keeping only relevant records for further analysis and building robust Machine Learning models. Data filtering in PySpark is pretty straightforward and can be done using two approaches:

1. filter

2. where

Data Filtering Using filter

Data filtering could be based on either a single column or multiple columns. The filter function in PySpark helps us to refine data based on specified conditions. If we were to filter data where the value of column "age" is less than 30, we could simply apply the filter function:

```
[In]: df.filter(df['age']<30).show()
[Out]:
```

▶ (1) Spark Jobs

```
+-------+---+----------+------+-------+
|ratings|age|experience|family| mobile|
+-------+---+----------+------+-------+
|      3| 27|      13.0|     3|  Apple|
|      4| 22|       2.5|     0|Samsung|
|      5| 27|       9.0|     1|     MI|
|      4| 27|       9.0|     0|   Oppo|
|      3| 22|       2.5|     0|  Apple|
|      3| 27|       6.0|     0|     MI|
|      2| 27|       6.0|     2|   Oppo|
|      5| 27|       6.0|     2|Samsung|
|      5| 27|       6.0|     0|     MI|
|      4| 22|       6.0|     1|   Oppo|
+-------+---+----------+------+-------+
only showing top 10 rows
```

As we can observe, we have all data columns as part of the filtered output for which the age column has values of less than 30. We can apply further filtering using "select" to print only specific columns. For example, if we want to view the ratings and mobile for customers whose age is less than 30, we can do that by using the select function after applying filter to the age column:

```
[In]: df.where(df['age']<30).select('ratings','mobile').show(10)
[Out]:
```

▸ (1) Spark Jobs

```
+-------+-------+
|ratings| mobile|
+-------+-------+
|      3|  Apple|
|      4|Samsung|
|      5|     MI|
|      4|   Oppo|
|      3|  Apple|
|      3|     MI|
|      2|   Oppo|
|      5|Samsung|
|      5|     MI|
|      4|   Oppo|
+-------+-------+
only showing top 10 rows
```

If data filtering is based on multiple columns, then we can do it in a couple of ways. We can either use sequential filtering in PySpark or make use of (&,|) operators to provide multiple filter conditions. Let us say we want customers with age less than 30 and who are only "Oppo" users. We apply filter conditions in sequence:

```
[In]: df.filter(df['age']<30).filter(df['mobile'] == 'Oppo').show()
[Out]:
```

```
+-------+---+----------+------+------+
|ratings|age|experience|family|mobile|
+-------+---+----------+------+------+
|      4| 27|       9.0|     0|  Oppo|
|      2| 27|       6.0|     2|  Oppo|
|      4| 22|       6.0|     1|  Oppo|
|      2| 27|       6.0|     2|  Oppo|
|      4| 22|       6.0|     1|  Oppo|
+-------+---+----------+------+------+
```

We can also use operators like & and | to apply multiple filter conditions and get the required data. In the following example, we filter for Oppo mobile users who have experience of greater than or equal to 9 using "&". In the follow-up example, we make use of the "or" condition to filter records with either "Oppo" users or "MI" users:

```
[In]: df.filter((df['mobile']=='Oppo')&(df['experience'] >=9)).show()
```

```
+-------+---+----------+------+------+
|ratings|age|experience|family|mobile|
+-------+---+----------+------+------+
|      4| 27|       9.0|     0|  Oppo|
|      2| 42|      23.0|     2|  Oppo|
|      2| 32|      16.5|     2|  Oppo|
+-------+---+----------+------+------+
```

```
[In]: df.filter((df['mobile']=='Oppo') | (df['mobile']=='MI')).show()
[Out]:
```

```
+-------+---+----------+------+------+
|ratings|age|experience|family|mobile|
+-------+---+----------+------+------+
|      5| 27|       9.0|     1|    MI|
|      4| 27|       9.0|     0|  Oppo|
|      3| 27|       6.0|     0|    MI|
|      2| 27|       6.0|     2|  Oppo|
|      5| 27|       6.0|     0|    MI|
|      4| 22|       6.0|     1|  Oppo|
|      1| 37|      23.0|     5|    MI|
|      2| 42|      23.0|     2|  Oppo|
|      3| 42|      23.0|     5|    MI|
|      5| 27|       2.5|     0|    MI|
|      2| 27|       6.0|     2|  Oppo|
|      2| 32|      16.5|     2|  Oppo|
|      3| 27|       6.0|     0|    MI|
|      3| 27|       6.0|     0|    MI|
|      4| 22|       6.0|     1|  Oppo|
+-------+---+----------+------+------+
```

In order to get a count of the number of records after filtering, count() can be used as shown in the following:

```
[In] : df.filter(df['age']<30).count()
```

Out[18]: 19

Data Filtering Using where

Data can also be filtered using the **where** function in PySpark. Most of the order remains similar to that of filter seen previously:

```
[In] : df.where(df['age']<30).count()
```

▶ (2) Spark Jobs

Out[18]: 19

```
[In]: df.where(df['age']<30).filter(df['mobile'] == 'Oppo').show()
```

```
+-------+---+----------+------+------+
|ratings|age|experience|family|mobile|
+-------+---+----------+------+------+
|      4| 27|       9.0|     0|  Oppo|
|      2| 27|       6.0|     2|  Oppo|
|      4| 22|       6.0|     1|  Oppo|
|      2| 27|       6.0|     2|  Oppo|
|      4| 22|       6.0|     1|  Oppo|
+-------+---+----------+------+------+
```

```
[In]: df.where((df['mobile']=='Oppo') | (df['mobile']=='MI')).show()
[Out]:
```

```
+-------+---+----------+------+------+
|ratings|age|experience|family|mobile|
+-------+---+----------+------+------+
|      5| 27|       9.0|     1|    MI|
|      4| 27|       9.0|     0|  Oppo|
|      3| 27|       6.0|     0|    MI|
|      2| 27|       6.0|     2|  Oppo|
|      5| 27|       6.0|     0|    MI|
|      4| 22|       6.0|     1|  Oppo|
|      1| 37|      23.0|     5|    MI|
|      2| 42|      23.0|     2|  Oppo|
|      3| 42|      23.0|     5|    MI|
|      5| 27|       2.5|     0|    MI|
|      2| 27|       6.0|     2|  Oppo|
|      2| 32|      16.5|     2|  Oppo|
|      3| 27|       6.0|     0|    MI|
|      3| 27|       6.0|     0|    MI|
|      4| 22|       6.0|     1|  Oppo|
+-------+---+----------+------+------+
```

We can add a new column in the Spark dataframe using the withColumn function in PySpark. For example, if we were to create a new column (new age) in the Dataframe by using the age column, we would simply use the withColumn function and add one year to the age column. This would result in a new column being created with updated values. This doesn't actually transform the Dataframe until it's assigned to the old Dataframe or a new Dataframe. So if we were to print our Dataframe again, we would observe that the new age column is missing as it wasn't assigned to the df Dataframe while creating it:

```
[In]: df.withColumn('new_age',df['age']+1).show(5)
```

▸ (1) Spark Jobs

```
+-------+---+----------+------+-------+-------+
|ratings|age|experience|family| mobile|new_age|
+-------+---+----------+------+-------+-------+
|      3| 32|       9.0|     3|   Vivo|     33|
|      3| 27|      13.0|     3|  Apple|     28|
|      4| 22|       2.5|     0|Samsung|     23|
|      4| 37|      16.5|     4|  Apple|     38|
|      5| 27|       9.0|     1|     MI|     28|
+-------+---+----------+------+-------+-------+
only showing top 5 rows
```

[In]: df.show()

```
+-------+---+----------+------+-------+
|ratings|age|experience|family| mobile|
+-------+---+----------+------+-------+
|      3| 32|       9.0|     3|   Vivo|
|      3| 27|      13.0|     3|  Apple|
|      4| 22|       2.5|     0|Samsung|
|      4| 37|      16.5|     4|  Apple|
|      5| 27|       9.0|     1|     MI|
|      4| 27|       9.0|     0|   Oppo|
|      5| 37|      23.0|     5|   Vivo|
|      5| 37|      23.0|     5|Samsung|
|      3| 22|       2.5|     0|  Apple|
|      3| 27|       6.0|     0|     MI|
|      2| 27|       6.0|     2|   Oppo|
|      5| 27|       6.0|     2|Samsung|
|      3| 37|      16.5|     5|  Apple|
|      5| 27|       6.0|     0|     MI|
|      4| 22|       6.0|     1|   Oppo|
|      4| 37|       9.0|     2|Samsung|
|      4| 27|       6.0|     1|  Apple|
```

If we assign it to a new dataframe (df_updated), then it would be present as an additional column as shown in the following:

```
[In]: df_updated=df.withColumn('new_age',df['age']+1)
[in]: df_updated.show(5)
```

```
+-------+---+----------+------+-------+-------+
|ratings|age|experience|family| mobile|new_age|
+-------+---+----------+------+-------+-------+
|      3| 32|       9.0|     3|   Vivo|     33|
|      3| 27|      13.0|     3|  Apple|     28|
|      4| 22|       2.5|     0|Samsung|     23|
|      4| 37|      16.5|     4|  Apple|     38|
|      5| 27|       9.0|     1|     MI|     28|
+-------+---+----------+------+-------+-------+
only showing top 5 rows
```

In terms of aggregating data with respect to individual columns, we can use the groupBy function in PySpark and cut the data based on various measures. For example, if we were to look at the number of users for every mobile brand, we would run a groupBy on "mobile" and take a count of users:

```
[In]: df.groupBy('mobile').count().show()
```

▶ (2) Spark Jobs

```
+-------+-----+
| mobile|count|
+-------+-----+
|     MI|    8|
|   Oppo|    7|
|Samsung|    6|
|   Vivo|    5|
|  Apple|    7|
+-------+-----+
```

The aggregate measure can be changed based on the requirement such as count, sum, mean, or min. In the following example, we run an aggregate sum of values in other columns for every mobile brand:

```
[In]: df.groupBy('mobile').sum().show()
```

```
+-------+------------+--------+----------------+------------+
| mobile|sum(ratings)|sum(age)|sum(experience)|sum(family)|
+-------+------------+--------+----------------+------------+
|     MI|          28|     241|            81.5|          11|
|   Oppo|          20|     199|            72.5|          10|
|Samsung|          25|     172|            52.0|          11|
|   Vivo|          21|     180|            57.0|           9|
|  Apple|          24|     214|            77.0|          19|
+-------+------------+--------+----------------+------------+
```

If we need different aggregate measures for specific columns, then we can make use of agg along with groupBy as shown in the following example:

```
[In]: df.groupBy('mobile').agg({'ratings':'mean'}).show()
```

```
+-------+------------------+
| mobile|      avg(ratings)|
+-------+------------------+
|     MI|               3.5|
|   Oppo| 2.857142857142857|
|Samsung| 4.166666666666667|
|   Vivo|               4.2|
|  Apple|3.4285714285714284|
+-------+------------------+
```

In order to find the distinct values in a Dataframe column, we can use the distinct function along with count as shown in the following:

```
[In]: df.select('mobile').distinct().count()
```

▶ (3) Spark Jobs

Out[35]: 5

In this part, we will understand UDFs – user-defined functions. Many times, we have to apply certain conditions and create a new column based on values in other columns of the Dataframe. In Pandas, we typically use the map or apply function to replicate such requirement. In PySpark, we can use UDFs to help us transform values in the Dataframe.

There are two types of UDFs available in PySpark:

1. Conventional UDF

2. Pandas UDF

Pandas UDFs are much more powerful in terms of execution speed and processing time. We will see how to use both types of UDFs in PySpark.

First, we need to import udf from PySpark and define the function that we need to apply to an existing Dataframe column. Now we can apply a basic UDF either by using a lambda or typical Python function. In this example, we define a custom Python function price range that returns if the mobile belongs to the High Price, Mid-Price, or Low Price category based on the brand. The next step is to declare the UDF and its return type (StringType in this example). Finally, we can use withColumn and mention the name of the new column to be formed along with applying the UDF and passing the relevant Dataframe column (mobile):

```
[In]: from pyspark.sql.functions import udf
[In]: from pyspark.sql.types import *
[In]: def price_range(brand):
  if brand in ['Samsung','Apple']:
    return 'High Price'
  elif brand =='MI':
    return 'Mid Price'
  else:
    return 'Low Price'
[In]: brand_udf=udf(price_range,StringType())
[In]: df.withColumn('price_range',brand_udf(df['mobile'])).show(10,False)
[Out]:
```

```
+-------+---+----------+------+-------+-----------+
|ratings|age|experience|family|mobile |price_range|
+-------+---+----------+------+-------+-----------+
|3      |32 |9.0       |3     |Vivo   |Low Price  |
|3      |27 |13.0      |3     |Apple  |High Price |
|4      |22 |2.5       |0     |Samsung|High Price |
|4      |37 |16.5      |4     |Apple  |High Price |
|5      |27 |9.0       |1     |MI     |Mid Price  |
|4      |27 |9.0       |0     |Oppo   |Low Price  |
|5      |37 |23.0      |5     |Vivo   |Low Price  |
|5      |37 |23.0      |5     |Samsung|High Price |
|3      |22 |2.5       |0     |Apple  |High Price |
|3      |27 |6.0       |0     |MI     |Mid Price  |
+-------+---+----------+------+-------+-----------+
only showing top 10 rows
```

As we can observe, a new column gets added to the DataFrame containing return values from the UDF being applied on original mobile column.

Pandas UDF

As mentioned before, Pandas UDFs are way faster and efficient compared with their peers. There are two types of Pandas UDFs:

1. Scalar

2. GroupedMap

Using a Pandas UDF is quite similar to using a basic UDF. We have to first import pandas_udf from pyspark.sql.functions and apply it on any particular column to be transformed:

```
[In]: from pyspark.sql.functions import pandas_udf
```

In this example, we define a Python function that calculates the number of years left in a user's life assuming life expectancy of total 100 years. It is a very simple calculation. We subtract the current age of the user from 100 using the Python function:

```
[In]:
def remaining_yrs(age):
  yrs_left=(100-age)
  return yrs_left
```

Once we create the pandas UDF (length_udf) using the Python function (remaining_yrs), we can apply it on the age column and create a new column yrs_left:

```
[In]: length_udf = pandas_udf(remaining_yrs, IntegerType())
[In]: df.withColumn("yrs_left", length_udf(df['age'])).show(10,False)
```

[Out]:

```
+-------+---+----------+------+-------+--------+
|ratings|age|experience|family|mobile |yrs_left|
+-------+---+----------+------+-------+--------+
|3      |32 |9.0       |3     |Vivo   |68      |
|3      |27 |13.0      |3     |Apple  |73      |
|4      |22 |2.5       |0     |Samsung|78      |
|4      |37 |16.5      |4     |Apple  |63      |
|5      |27 |9.0       |1     |MI     |73      |
|4      |27 |9.0       |0     |Oppo   |73      |
|5      |37 |23.0      |5     |Vivo   |63      |
|5      |37 |23.0      |5     |Samsung|63      |
|3      |22 |2.5       |0     |Apple  |78      |
|3      |27 |6.0       |0     |MI     |73      |
+-------+---+----------+------+-------+--------+
only showing top 10 rows
```

The preceding example showed how to use a Pandas UDF on a single dataframe column, but sometimes we might need to use more than one column. Hence, let's go over one more example where we understand the method of applying a Pandas UDF on multiple columns of a dataframe. Here we will create a new column, which is simply the product of ratings and experience columns. As usual, we define a Python function and calculate the product of the two columns:

```
[In]:
def prod(rating,exp):
  x=rating*exp
  return x
[In]: prod_udf = pandas_udf(prod, DoubleType())
```

After creating the Pandas UDF, we can apply it on both of the columns (ratings, experience) to form the new column (product):

```
[In]: df.withColumn("product",prod_udf(df['ratings'],df['experience'])).
show(10, False)
[Out]:
```

```
+-------+---+----------+------+-------+-------+
|ratings|age|experience|family|mobile |product|
+-------+---+----------+------+-------+-------+
|3      |32 |9.0       |3     |Vivo   |27.0   |
|3      |27 |13.0      |3     |Apple  |39.0   |
|4      |22 |2.5       |0     |Samsung|10.0   |
|4      |37 |16.5      |4     |Apple  |66.0   |
|5      |27 |9.0       |1     |MI     |45.0   |
|4      |27 |9.0       |0     |Oppo   |36.0   |
|5      |37 |23.0      |5     |Vivo   |115.0  |
|5      |37 |23.0      |5     |Samsung|115.0  |
|3      |22 |2.5       |0     |Apple  |7.5    |
|3      |27 |6.0       |0     |MI     |18.0   |
+-------+---+----------+------+-------+-------+
only showing top 10 rows
```

Drop Duplicate Values

We can use the dropDuplicates function in order to remove the duplicate records from the Dataframe. The total number of records in this Dataframe is 33, but it also contains 7 duplicate records, which can easily be confirmed by dropping those duplicate records as we are left with only 26 rows:

```
[In]: df.count()
[Out]: 33
[In]: df=df.dropDuplicates()

[In]: df.count()
[Out]: 26
```

In order to delete a column or multiple columns, we can use the drop functionality in PySpark:

```
[In]: df.drop('age').show()
```

```
+-------+----------+------+-------+
|ratings|experience|family| mobile|
+-------+----------+------+-------+
|      3|       9.0|     3|   Vivo|
|      3|      13.0|     3|  Apple|
|      4|       2.5|     0|Samsung|
|      4|      16.5|     4|  Apple|
|      5|       9.0|     1|     MI|
+-------+----------+------+-------+
only showing top 5 rows
```

```
[In]: df.drop('age','experience').show(5)
[Out]:
```

```
+-------+------+-------+
|ratings|family| mobile|
+-------+------+-------+
|      3|     3|   Vivo|
|      3|     3|  Apple|
|      4|     0|Samsung|
|      4|     4|  Apple|
|      5|     1|     MI|
+-------+------+-------+
only showing top 5 rows
```

Writing Data

Once we have the processing steps completed, we can write the clean Dataframe to a desired location (local/cloud) in the required format. In our case, we simply write back the sample processed data to Databricks FileStore. We can choose the desired format (.csv, .parquet, etc.) to save the final Dataframe.

CSV

If we want to save it back in the original .csv format as a single file, we can also use the coalesce function in Spark:

```
[In]:df.write.format('csv').option('header','true').save("/FileStore/
tables/processed_sample_data.csv")
```

We can append the file name with v1 to avoid conflict due to the same file name as processed_sample_data might already exist:

```
[In]:df.coalesce(1).write.format('csv').option('header','true').save("/
FileStore/tables/processed_sample_data_v1.csv")
```

Parquet

If the dataset is huge and involves a lot of columns, we can choose to compress it and convert it into a .parquet file format. It reduces the overall size of the data and optimizes the performance while processing data because it works on a subset of required columns instead of the entire data. We can convert and save the Dataframe into the .parquet format easily by mentioning the format as parquet as shown in the following:

```
[In]:df.write.format('parquet').option('header','true').save("/FileStore/
tables/processed_sample_data.csv")
```

Data Handling Using Koalas

Data scientists and data engineers often start their data wrangling journey with Pandas as it is the most established data processing library out there along with NumPy and scikit-learn for Machine Learning. Given the adoption rate of Pandas is so high, there is no debate around the use of Pandas as a standard data processing library. It is quite mature in terms of offerings, simple APIs, and extensibility. However, one of the key limitations of using Pandas is that you can hit a roadblock if the data you're dealing with is very big in terms of size as Pandas is not very scalable. The core reason is Pandas is meant to run on a single-worker machine instead of a multiple-worker setup. Hence, for large datasets, Pandas will either run incredibly slow or usually throw memory errors. However, for proof of concepts (POCs) and minimum viable products (MVPs), Pandas is still the go-to library. The other libraries such as Dask try to address the scalability issue that Pandas has by processing data in partitions and speeding up things in the execution front.

There are some challenges that still persist. Hence, when it comes to handling big data and distributed processing frameworks, Spark becomes the de facto choice for the data community out there. One of the new offerings by Databricks is the open source library called Koalas. It provides a Pandas DataFrame API on top of Apache Spark. As a result, it takes advantage of the Spark implementation of DataFrames, query optimization, and data source connectors all with Pandas syntax. In essence, it offers a nice alternative to use the power of Spark even while working with Pandas.

Note   This Koalas library is still under active development and covering more than 75% of the Pandas API.

There are a number of benefits of using Koalas over Pandas when dealing with large datasets as it allows to use similar syntax as that of Pandas without worrying too much about the underlying Spark details. One can easily transition between Pandas and Koalas Dataframes and into Spark Dataframes. It also offers easy integration with SQL through which one can run multiple queries on a Koalas Dataframe.

The first step is to import koalas and convert a Spark Dataframe to a Koalas Dataframe:

```
[In]: from databricks import koalas as ks
[In]: k_df = df.to_koalas()
```

If we look at the type of the new Dataframe, it would return it as a Koalas Dataframe type:

```
[In]: type(k_df)
[Out]: databricks.koalas.frame.DataFrame
```

Now we can simply start using the Pandas functions and syntax to transform and process the dataset. For example, a simple df.head(3) would return the top three rows of the dataframe:

```
[In]: k_df.head(3)
```

	ratings	age	experience	family	mobile
0	3	32	9.0	3	Vivo
1	3	27	13.0	3	Apple
2	4	22	2.5	0	Samsung

We can also look at the shape of the dataframe now using the shape function, which isn't possible in PySpark directly:

```
[In]: k_df.shape
```

▶ (2) Spark Jobs

Out[10]: (33, 5)

Finally just to reinforce, we can also use groupby, instead of groupBy, on a Koalas Dataframe to aggregate values based on different columns:

```
[In]: k_df.groupby(['mobile']).sum()
```

mobile	ratings	age	experience	family
MI	28	241	81.5	11
Oppo	20	199	72.5	10
Samsung	25	172	52.0	11
Vivo	21	180	57.0	9
Apple	24	214	77.0	19

Conclusion

In this chapter, we looked at the steps to read, explore, process, and write data using PySpark. We also looked at Koalas, which offers Pandas users to process data using Spark under the hood.

CHAPTER 3

Introduction to Machine Learning

When we are born, we are literally incapable of doing anything. We can't even hold our head straight for a few months, but eventually, we start learning. During those days, we all fumble, make tons of mistakes, fall down, and bang our head many times but slowly learn our way to sit, walk, speak, and finally run. As a built-in mechanism, we don't require a lot of examples to learn about things around us. For example, just by observing two to three dogs around, we can easily learn to recognize a dog from a cat. We can easily differentiate between a cycle and a car by seeing a few cars and bikes around. Even though it seems very easy and intuitive to us as human beings, for machines it can be a herculean task.

Machine learning is the mechanism through which we try to make machines learn without explicitly programming them to do so. For example, we expose the machine to a lot of photos of cats and dogs, just enough for the machine to learn the difference between the two and predict on unseen photos correctly. The question here might be that why do we need so many photos to learn the difference between cats and dogs. The challenge that the machines face is that they're not able to learn the entire pattern or abstraction features just from a few images. They would need enough examples (different in some way) to learn as much features as possible to be able to make right predictions, whereas as humans we have this amazing ability to draw abstraction at different levels and easily recognize objects. This example might be specific to the image recognition case, but for other applications as well, the machine would need a good amount of data to learn from before being able to predict on unseen data with reasonable accuracy.

There is no denying the fact that the world has seen significant progress in terms of machine learning and AI applications in the last decade or so. In fact, if they are to be compared with any other technology, ML and AI have been breaking paths in

© Pramod Singh 2022
P. Singh, *Machine Learning with PySpark*, https://doi.org/10.1007/978-1-4842-7777-5_3

multiple ways. Businesses such as Amazon, Apple, Google, and Facebook are thriving on these advancements in AI and are partly responsible to them as well. The research and development wings of organizations like these are pushing the limits and making incredible progress in bringing AI to everyone. Not only big names like these but thousands of start-ups have emerged on the landscape specializing in AI-based products and services. The numbers only continue to grow as I write this chapter. As mentioned earlier, the adoption of ML and AI by various businesses has exponentially grown over the last decade due to the following core reasons:

1. Rise in data

2. Increased computational efficiency

3. Improved ML algorithms

4. Availability of data scientists

Rise in Data

The first most prominent reason for this trend is the massive rise in data generation in the past couple of decades. Every device is generating data these days, as well as mobile apps, online shopping behavior, and server logs. There is so much data generated that there is a huge demand of people who can process and analyze data.

Increased Computational Efficiency

We have to understand the fact that ML and AI at the end of the day are simply dealing with a huge set of numbers being put together and made sense out of. To apply ML or AI, there is a heavy need of powerful processing systems, and all of us have witnessed significant improvements in computational power at a neckbreaking pace. Just to observe the changes that we have seen in the last decade or so, the size of mobiles has reduced drastically, and speed has increased to a great extent. This is not just in terms of physical changes in the microprocessor chips for faster processing using GPUs and TPUs but also the presence of data processing frameworks such as Spark. The combination of advancement in processing capabilities and in-memory computations using Spark made it possible for a lot of ML algorithms to be able to run successfully in the last decade.

Improved ML Algorithms

Over the last few years, there has been a tremendous progress in terms of availability of new and upgraded algorithms that have not only improved the prediction accuracy but also solved multiple challenges that traditional ML faced. In the first phase, which was rule-based systems, one had to define all the rules first and then design the system within that set of rules. It became increasingly difficult to control and update the number of rules as the environment was too dynamic. Hence, traditional ML came into the picture to replace rule-based systems. The challenge with this approach was that the data scientist had to spend a lot of time to hand-design the features for building the model (known as feature engineering) and there was an upper threshold in terms of prediction accuracy that these models could never go above no matter if the input data size was increased. The third phase was introduction of deep neural networks where a network would figure out the most important features on its own and also outperform other ML algorithms. Apart from these, some other approaches that are creating a lot of buzz over the last few years are

1. Meta-learning

2. Transfer leaning (nanonets)

3. Capsule networks

4. Deep reinforcement learning

5. Generative adversarial networks (GANs)

Availability of Data Scientists

ML and AI are a specialized field as there are multiple skills required to play this role well. To be able to build and apply ML models, one needs to have a sound knowledge of math and statistics fundamentals along with a good understanding of Machine Learning algorithms and various optimization techniques. The next important skill is to be extremely comfortable at coding to package your code for production grade. There is a huge excitement in the job market with respect to data scientist roles; and there are a huge number of requirements for data scientists everywhere especially in regions like the USA, the UK, and India.

As mentioned previously, Machine Learning has got a lot of attention in the last few years. More and more businesses want to adopt it to maintain the competitive edge. However, very few really have the right resources and the appropriate data to implement it. In this chapter, we will cover basic types of Machine Learning and how businesses can benefit from using Machine Learning. Those who are already aware of basic concepts and applications of Machine Learning can feel free to jump to the next chapter directly.

There are tons of definitions of Machine Learning on the Internet. However, if I tried to put in simple terms, it would look something like this:

> *Machine Learning is using statistical techniques and sometimes advanced algorithms to either make predictions or learn hidden patterns within the data and essentially replacing rule-based systems to make data-driven systems more powerful.*

Let's go through this definition in some more detail. Machine learning as the name suggests makes a machine learn, although there are many components that come into the picture when we talk about making a machine learn.

The first one is data that is the backbone for any sort of model. Machine learning thrives on relevant data. The more signals present in the data, the better the predictions. Machine Learning can be applied in different domains such as finance, retail, healthcare, and manufacturing. The other part is the underlying algorithm. Based on the nature of the problem, we choose the algorithm accordingly. Some algorithms do a better job compared with others in a particular context. The last part is the hardware and software aspect. The availability of open source distributed computing frameworks like Spark and TensorFlow has made Machine Learning tools easily accessible to more people. The rule-based systems came into the picture when the scenarios were limited and all the rules could be configured manually to handle the situations. For example, the manner in which a fraud can happen has dramatically changed over the past few years, and hence, creating manual rules for preventing such incident can be very difficult, whereas Machine Learning can be leveraged in such a scenario where the model learns from the data and adapts itself to the new data to make correct decisions accordingly.

Let's look at the different types of machine learning and their applications. We can categorize machine learning into four major categories as shown in Figure 3-1:

1. Supervised Machine Learning

2. Unsupervised Machine Learning

3. Semi-supervised Machine Learning

4. Reinforcement Learning

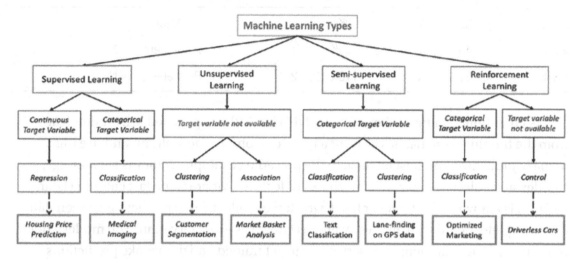

Figure 3-1. *Machine Learning categories*

Each of the preceding categories is used for specific purposes, and the data that is used also differs from each other. At the end of the day, machine learning is learning from data (historical or real time) and making decisions (offline or real time) based on the model training.

Supervised Machine Learning

This is the prime category of machine learning, which drives a lot of applications and value for businesses. In supervised learning, we train our models on labeled data. By labeled, it means to have the correct answers or outcome for the data. Let's take an example to illustrate supervised learning. If there is a financial company that wants to filter customers based on their profiles before accepting their loan request, the machine learning model would get trained on historical data that contains information regarding profiles of past customers and the label column on whether a customer has defaulted on loan or not. The sample data looks as given in Table 3-1.

Table 3-1. *Sample Customer base*

Customer ID	Age	Gender	Salary	No. of Loans	Job Type	Loan Default
AL23	32	M	80K	3	Contract	Y
AX43	45	F	150K	1	Permanent	N
BG76	51	M	110K	2	Permanent	N

As I mentioned in my earlier version of this book, the supervised ML models learn from the training data that has also got a label/outcome/target column and use this to make predictions on unseen data. In the preceding example, columns such as Age, Gender, and Salary are known as attributes or features, whereas the last column (Loan Default) is known as the target or label, which the model tries to predict for unseen data. One complete record with all these values is known as an observation. The model would require a sufficient amount of observations to get trained and then make predictions on a similar kind of data. There needs to be at least one input feature/attribute for the model to get trained along with the output column in supervised learning. The reason the machine is able to learn from the training data is that some of these input features individually or in combination have an impact on the output column (Loan Default).

There are many applications that use supervised learning settings such as

1. If a particular customer would buy the product or not

2. If the visitor would click the ad or not

3. If the person would default on loan or not

4. What is the expected sale price of a given property?

These are some of the applications of supervised learning, and there are many more. The methodology that is used sometimes varies based on the kind of output the model is trying to predict. If the target feature is a categorical type, then its falls under the classification category; and if the target feature is a numerical value, it would fall under the regression category. Some of the supervised ML algorithms are

1. Linear regression (LR)

2. Logistic regression

3. Support vector machines

4.   Naive Bayesian classifier

5.   Decision trees

6.   Ensemble methods

Another property of supervised learning is that the model's performance can be evaluated. Based on the type of model (classification/regression/time-series), the evaluation metric can be applied, and performance can be measured. This happens mainly by splitting the training data into two sets (train set and validation set) and training the model on the train set and testing its performance on the validation set since we already know the right label/outcome for the validation set. We can then make the changes in the hyperparameters (covered in later chapters) or introduce new features using feature engineering to improve the performance of the model.

Unsupervised Machine Learning

This is another category of machine learning that is used heavily in business applications. It is different from supervised learning in terms of the output labels. In unsupervised learning, we build the models on a similar sort of data as that of supervised learning except for the fact that this dataset does not contain any label or outcome column. Essentially, we apply the model on data without any right answers. In unsupervised learning, the machine tries to find hidden patterns and useful signals in the data, which can be later used for other applications. The main objective is to probe the data and come up with hidden patterns and similarity structure within the dataset as shown in Figure 3-2. One of the use cases is to find patterns within customer data and group the customers into different clusters. It can also identify those attributes that distinguish between any two groups. From a validation perspective, there is no measure of accuracy for unsupervised learning. The clustering done by person A can be totally different from that of person B based on parameters used to build the model. There are different types of unsupervised learning:

1.   K-means clustering

2.   Mapping of nearest neighbor

Clustering

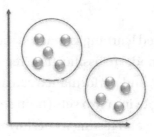

Customer Segmentation

Figure 3-2. *Clustering*

Semi-supervised Learning

As the name suggests, semi-supervised learning lies somewhere in between both supervised and unsupervised learning. In fact, it uses both of the techniques. This type of learning is mainly relevant in scenarios where we are dealing with a mixed sort of dataset, which contains both labeled and unlabeled data. Sometimes it's just unlabeled data completely, but we label some part of it manually. The whole idea of semi-supervised learning is to use this small portion of labeled data to train the model and then use it for labeling the other remaining part of data, which can then be used for other purposes. This is also known as pseudo-labeling as it labels the unlabeled data using the predictions made by the supervised model. To quote a simple example, let's say we have a lot of images of different brands from social media, and most of them are unlabeled. Now using semi-supervised learning, we can label some of these images manually and then train our model on the labeled images. We then use the model predictions to label the remaining images to transform the unlabeled data to labeled data completely.

The next step in semi-supervised learning is to retrain the model on the entire labeled dataset. The advantage that it offers is that the model gets trained on a bigger dataset, which was not the case earlier, and the model is now more robust and better at predictions. The other advantage is that semi-supervised learning saves a lot of effort and time, which could go into manually labeling the data. The flipside of doing all this is that it's difficult to get high performance of the pseudo-labeling as it uses a small part of labeled data to make the predictions. However, it's still a better option rather than manually labeling the data, which can be very expensive and time consuming at the same time. This is how semi-supervised learning uses both supervised and unsupervised learning to generate the labeled data. Businesses that face challenges regarding costs associated with the labeled training process usually go for semi-supervised learning.

Reinforcement Learning

This is the fourth kind of learning and is a little different in terms of data usage and its predictions. Reinforcement learning is a big research area in itself, and an entire book can be written just on it. The main difference between the other kinds of learning and reinforcement learning is that in the other kinds of learning, we need data, mainly historical data, to train the models, whereas reinforcement learning works in a reward system as shown in Figure 3-3. It is primarily decision-making based on certain actions that the agent takes to change its state in order to maximize the rewards. Let's break this down to individual elements using a visualization.

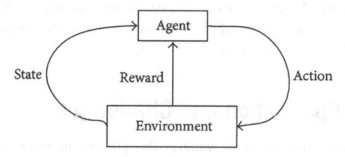

Figure 3-3. *Reinforcement learning*

1. Autonomous Agent: This is the main character in this whole learning who is responsible to take action. If it is a game, the agent makes the moves to finish or reach the end goal.

2. Actions: These are a set of possible steps the agent can take in order to move forward in the task. Each action will have some effect on the state of the agent and can result in either reward or penalty. For example, in a game of tennis, actions might be to serve, return, move left or right, etc.

3. Reward: This is the key to make the progress in reinforcement learning. The agent takes actions that can result in either reward or penalty. It is an instant feedback mechanism, which differentiates it from traditional supervised and unsupervised learning techniques.

4. Environment: This is the territory in which the agent gets to play in. The environment decides whether the actions the agent takes result in reward or penalty.

5. State: The position the agent is in at any given point in time defines the state of the agent. To move forward or reach the end goal, the agent has to keep changing states in the positive direction to maximize the rewards.

The unique thing about reinforcement learning is that there is an immediate feedback mechanism that drives the next behavior of the agent based on the reward system. Most of the applications that use reinforcement learning are in navigation, robotics, and gaming. However, it can be also used to build recommender systems. Now that we have a basic understanding of Machine Learning, let's look at different applications of ML in various domains.

Industrial Application and Challenges

In the final section of this chapter, we would go though some of the real applications of ML and AI. Businesses are heavily investing in ML and AL across the globe and establishing standard procedures to leverage capabilities of ML and AI to build their competitive edge. There are multiple areas where ML and AI are being currently applied and providing great values for the businesses. We will look at a few of the major domains where ML and AI are transforming the landscape.

Retail

One of the business verticals that is making incredible use of ML and AI is retail. Since retail business generates a lot of customer data, it offers a perfect platform for applying ML and AI. The retail sector has always faced multiple challenges such as a stock-out situation, suboptimal pricing, limited cross-sell or upsell, and inadequate personalization. ML and AI have been able to address many of these challenges and offered incredible impact in the retail space. There have been numerous applications built in the retail space that are powered by ML and AI in the last decade, and it continues to grow at a neckbreaking pace. The most prominent application is the recommender system. Online retail businesses are thriving on recommender systems as these systems can increase their revenue by a great deal. Apart from recommender systems, retail uses ML and AI capabilities for stock optimization to control inventory levels and reduce costs. Dynamic pricing is another area where AI and ML are being used comprehensively to get maximum returns. Customer segmentation is also done using ML as it uses not only demographics information of customers but also transactional data and takes multiple other variables into consideration before revealing the different groups within the customer base. Product categorization is also being done using ML as it saves huge manual effort and increases the accuracy level of labeling the products. Demand forecasting and stock optimization are tackled using ML and AI in order to save costs. Route planning has also been handled by ML and AI in the last few years as it enables businesses to fulfill orders in a more effective way. As a result of ML and AI applications in retail, cost savings have improved, businesses are able to take informed decisions, and overall customer satisfaction has gone up.

Healthcare

Another business vertical to be deeply impacted by ML and AI is healthcare. Diagnosis based on image data using ML and AI is being adopted at a quick rate across the healthcare spectrum. The prime reason is the level of accuracy offered by ML and AI and the ability to learn from data of the past decades. ML and AI algorithms on X-rays, MRI scans, and various other images in the healthcare domain are being heavily used to detect any anomalies. A virtual assistant or chatbot is also being deployed as a part of applications to assist with explaining lab reports. Finally, insurance verification is also being done using ML models in healthcare to avoid any inconsistency.

Finance

The finance domain has always had data, lots of it. Out of all the domains, finance always has been data enriched. Hence, there have been multiple applications built over the last decade based on ML and AI. The most prominent one is the fraud detection system that uses anomaly detection algorithms in the background. Other areas are portfolio management and algorithmic trading. ML and AI have the ability to scan over 100 years of past data and learn the hidden patterns in order to suggest the best calibration to a portfolio. Complex AI systems are being used to make extremely fast decisions on trading to maximize the gains. ML and AI are also used in risk mitigation and loan insurance underwriting. Again, recommender systems are used to upsell and cross-sell various financial products by various institutions. They also use them for predicting the churn of the customer base in order to formulate a strategy to retain the customers who are likely to discontinue with a specific product or service. Another important usage of ML and AI in the finance sector is to check if the loan should be granted or not to various applicants based on predictions made by the model. Apart from that, ML is being used to validate if the insurance claims are genuine or fraud based on the ML model predictions.

Travel and Hospitality

Just as retail, the travel and hospitality domain is thriving on ML- and AI-based applications. To name a few, recommender systems, price forecasting, and virtual assistants are all ML- and AI-based applications that are being leveraged in travel and hospitality verticals. Starting from recommending best deals to alternative travel dates, recommender systems are supercritical to drive customer behavior in this sector. It also recommends new travel destinations based on user preferences, which are highly tailored using ML in the background. AI is also being used to send timely alerts to customers by predicting future price movements based on various factors. Virtual assistants nowadays are part of every travel website as customers don't want to wait to get relevant information. On top of it, interactions with these virtual assistants are very humanlike as natural language understanding intelligence is already embedded into these chatbots to a great extent so as to understand simple questions and reply in a similar manner.

Media and Marketing

Every business more or less depends on marketing in order to get more customers, and reaching out to the right customer has always been a big challenge. Thanks to ML and AI, that problem is now better handled as the technology can anticipate customer behavior to a great extent. ML- and AI-based applications are being used to differentiate between potential prospects who are more likely to buy or subscribe to the offer or product and casual candidates. They're also being used to provide an absolute personalized offer in order to convert or retain customers. A churn predictor is again used heavily to identify the group of consumers who are likely to discontinue the usage of any particular product or service. Advanced customer segmentation for hypertargeting is being done using ML and AI. Finally, a lot of marketing content is being generated artificially using ML and AI these days in order to send out the best-performing content.

Manufacturing and Automobile

The manufacturing domain has not been able to escape the wave of ML and AI as well. The most predominant one is predictive maintenance as ML- and AI-based applications can help in preventing potential damages by predicting the need of maintenance in advance based on earlier data. Automobile companies use telematics data in order to learn the driving patterns of the customers and act more promptly to help them in many ways. They also use web data to understand their customers better to try and personalize the experience for seamless navigation during the online journey.

Social Media

Most of the people out there, the young generation in particular, spend a great deal of time on social media without realizing the fact that a lot of the applications are using ML and AI. Facebook, YouTube, LinkedIn, Twitter, and other similar apps use ML heavily in providing the experience. Right from photo auto-tag suggestion to recommendation of friends, everything is driven by ML and AI. They are also used to generate subtitles and language translations for various platforms such as YouTube. Various search engines and voice assistants are using good amount of ML implementation in them.

Others

There are many other applications where ML and AI are used. For example, email spam filtering nowadays uses ML instead of a rule-based system. One advantage the ML approach offers over the traditional rule-based system is that the former automatically updates and upgrades itself as per the new mails to make this distinction. Another area is the oil and gas industry where ML and AI help in analyzing underground minerals and finding alternative energy sources. They are also being used in transportation as they can predict likely traffic conditions and alert you in advance.

Conclusion

In this chapter, we went over the fundamentals of Machine Learning and also covered the different applications of Machine Learning along with its existing challenges.

CHAPTER 4

Linear Regression

As we talked about in the previous chapter, Machine Learning is a very vast field, and there are multiple algorithms that fall under various categories, but linear regression is one of the most fundamental Machine Learning algorithms. This chapter focuses on building a linear regression model with PySpark and dives deep into the workings of a LR model. It will cover various assumptions to be considered before using LR along with different evaluation metrics. But before even jumping into understanding linear regression, we must understand types of variables.

Variables

Variables capture data information in different forms. There are mainly two categories of variables that are used widely as depicted in Figure 4-1.

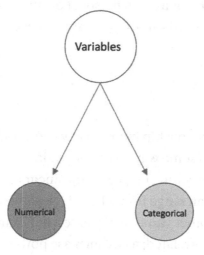

Figure 4-1. *Variable types*

© Pramod Singh 2022
P. Singh, *Machine Learning with PySpark*, https://doi.org/10.1007/978-1-4842-7777-5_4

We can even further break down these variables into subcategories, but we will stick to these two types throughout this book. Numerical variables are those kinds of values that are quantitative in nature such as numbers (integers/floats). For example, salary records, exam scores, age or height of a person, and stock prices all fall under the category of numerical variables.

Categorical variables on the other hand are qualitative in nature and mainly represent categories of data being measured, for example, colors, outcome (yes/no), and ratings (good/poor/average).

For building any sort of machine learning model, we need to have input and output variables. Input variables are those values that are used to build and train the machine learning model to predict the output or target variable. Let's take a simple example. Suppose we want to predict the salary of a person given the age of the person using machine learning. In this case, the salary is our output/target/dependent variable as it depends on age, which is known as the input or independent variable. Now the output variable can be categorical or numerical in nature, and depending on its type, machine learning models are chosen.

Now coming back to linear regression, it is primarily used in cases where we are trying to predict a numerical output variable. Linear regression is used to predict a line that fits the input data points the best possible way and can help in predictions for unseen data, but the point to notice here is that how can a model learn just from "age" and predict the salary amount for a given person. For sure, there needs to be some sort of relationship between these two variables (salary and age). There are two major types of variable relationships:

- Linear

- Nonlinear

The notion of a linear relationship between any two variables suggests that both are proportional to each other in some ways. The correlation between any two variables gives us an indication on how strong or weak is the linear relationship between them. The correlation coefficient can range from –1 to +1. Negative correlation means as one of the variables increases, the other variable decreases. For example, power and mileage of a vehicle can be negatively correlated; as we increase power, the mileage of the vehicle goes down. On the other hand, salary and years of work experience are an example of positively correlated variables. Nonlinear relationships are comparatively complex in nature and hence require an extra amount of details to predict the target variables. For example, in a self-driving car, the relationship between input variables such as terrain, signal system, and pedestrian to the speed of the car is nonlinear.

Note   The next section includes theory behind linear regression and might be redundant for many readers. Please feel free to skip the section in that case.

Theory

Now that we understand basics of variables and relationships between variables, let's build on the example of age and salary to understand linear regression in depth.

The overall objective of linear regression is to predict a straight line through the data such that the vertical distance of each of these points is minimal from that line. So, in this case, we will predict salary of a person given the age. Let's assume we have records of four people, which capture age and their respective salaries, as shown in Table 4-1.

Table 4-1. *Sample Data*

Sr. No	Age	Salary ('0000 $)
1	20	5
2	30	10
3	40	15
4	50	22

We have an input variable (age) at our disposal to make use of in order to predict the salary (which we will do at a later stage in this book), but let's take a step back. Let's assume all we have with us at the start is just the salary values of these four people. The salary is plotted for each person in Figure 4-2.

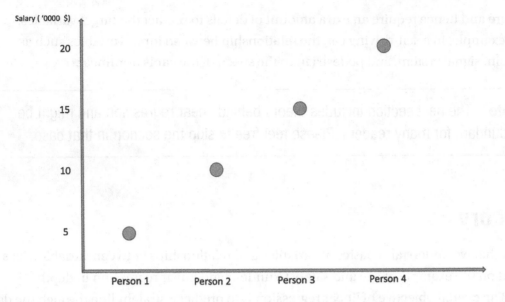

Figure 4-2. *Scatter plot of salary*

Now, we if we were to predict the salary of the fifth person (new person) based on the salaries of these earlier people, the best possible way to predict that would be to take an average/mean of the existing salary values. That would be the best prediction given this information. It is like building a Machine Learning model but without any input data (since we are using the output data as input data).

Let's go ahead and calculate the average salary for these given salary values:

$$\text{Avg. Salary} = \frac{(5+10+15+22)}{4} = \textbf{13}$$

So the best prediction of the salary value for the next person is 13. Figure 4-3 showcases the salary values for each person along with the mean value (the best-fit line in the case of using only one variable).

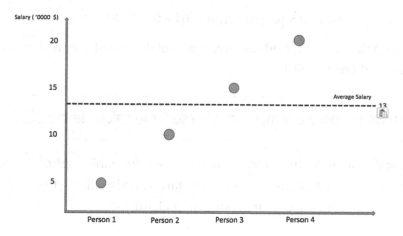

Figure 4-3. *Best-fit line plot*

The line for the mean value as shown in Figure 4-3 is possibly the best-fit line in this scenario for these data points because we are not using any other variable apart from salary itself. If we take a look closely, none of the earlier salary values lies on this best-fit line. There seems to be some amount of separation from the mean salary value as shown in Figure 4-4. These are also known as errors. If we go ahead and add them up and calculate the total sum of this distance, it becomes 0, which makes sense since it's the mean value of all the data points. So, instead of simply adding them, we square each error and then add them up.

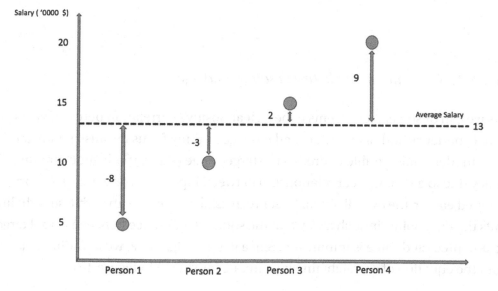

Figure 4-4. *Residuals plot*

$$\text{Sum of Squared Errors} = 64 + 9 + 4 + 81 = 158$$

So adding up the squared residuals gives us a total value of 158, which is known as the sum of squared errors (SSE).

Note   We have not used any input variable so far to calculate the SSE.

Let us park this score for now and include the input variable (age of the person) as well to predict the salary of the person. Let's start with visualizing the relationship between age and salary of the person as shown in Figure 4-5.

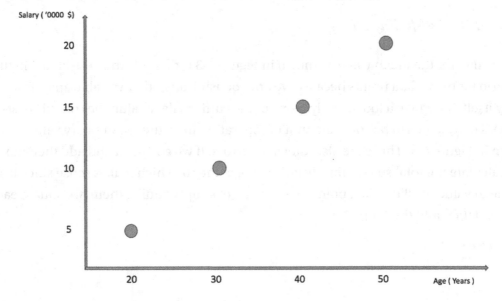

Figure 4-5. *Correlation plot between salary and age*

As we can observe, there seems to be a clear positive correlation between years of work experience and salary value, and it is a good thing for us because it indicates that the model would be able to predict the target value (salary) with good amount of accuracy due to a strong linear relationship between input (age) and output (salary). As mentioned earlier, the overall aim of linear regression is to come up with a straight line that fits the data points in such a way that the squared difference between actual target value and predicted value is minimized. Since it is a straight line, we know in linear algebra the equation of a straight line is y= mx + c (as shown in Figure 4-6).

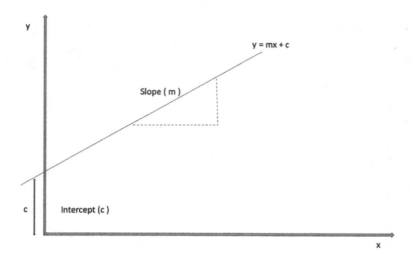

Figure 4-6. *Straight line plot*

where

m = slope of the line $\left(\dfrac{x_2 - x_1}{y_2 - y_1} \right)$

x = value at x-axis

y= value at y-axis

c = intercept (value of y at x = 0)

Since linear regression is also finding out the straight line, the linear regression equation becomes

$$y = B_0 + B_1 * x$$

(since we are using only one input variable, i.e., age)

where

y = salary (prediction)

B0 = intercept (value of salary when age is 0)

B1 = slope or coefficient of salary

x= age

Now, you may ask that there can be multiple lines that can be drawn through the data points (as shown in Figure 4-7) and how to figure out which is the best-fit line.

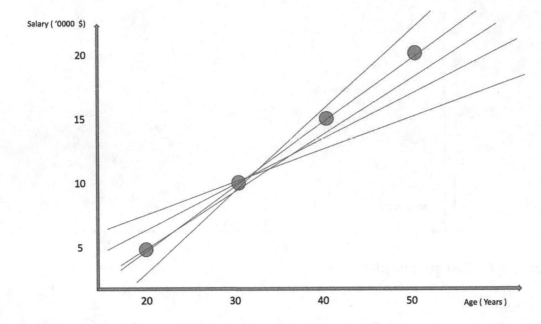

Figure 4-7. *Possible straight lines through data*

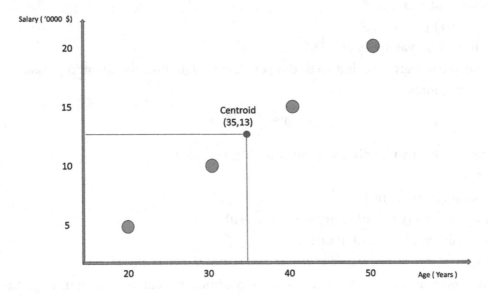

Figure 4-8. *Centroids of data*

The first criterion to find out the best-fit line is that it should pass through the centroids of the data points as shown in Figure 4-8. In our case, the centroid values are

$$\text{mean (Age)} = \frac{(20+30+40+50)}{4}$$
$$= 35$$
$$\text{mean (Salary)} = \frac{(5+10+15+22)}{4}$$
$$= 13$$

The second criterion is that it should be able to minimize the sum of squared errors. We know our regression line equation is equal to

$$y = B_0 + B_1 * x$$

Now the objective of using linear regression is to come up with the most optimal values of the intercept (B_0) and coefficient (B_1) so that the residuals/errors are minimized to the maximum extent.

We can easily find out the values of B_0 and B_1 for our dataset by using the following formulas:

$$B_1 = \frac{\sum (x_i - x_{mean}) * (y_i - y_{mean})}{\sum (x_i - x_{mean})^2}$$

$$B_0 = y_{mean} - B_1 * (x_{mean})$$

Table 4-2 showcases the calculation of slope and intercept for linear regression using input data.

Table 4-2. *Calculation of Slope and Intercept*

Age	Salary	Age Variance (Diff. from Mean)	Salary Variance (Diff. from Mean)	Covariance (Product)	Age Variance (Squared)
20	5	-15	-8	120	225
30	10	-5	-3	15	25
40	15	5	2	10	25
50	22	15	9	135	225

Mean (Age) = 35
Mean (Salary) =13

The covariance between any two variables (age and salary) is defined as the product of the distances of each variable (age and salary) from their mean. In short, the product of the variances of age and salary is known as covariance. Now that we have the covariance and age variance squared values, we can go ahead and calculate the values of slope and intercept of the linear regression line:

$$B_1 = \frac{\sum(Covariance)}{\sum(Age\ Variance\ Squared)}$$

$$= \frac{280}{500}$$

$$= 0.56$$

$$B_0 = 13 - (0.56 * 35)$$

$$= -6.6$$

Our final linear regression equation becomes

$$y = B_0 + B_1 * x$$

$$Salary = -6.6 + (0.56 * Age)$$

We can now predict any of the salary values using this equation given any age. For example, the model would have predicted the salary of the first person something like this:

$$Salary\ (1st\ person) = -6.6 + (0.56 * 20)$$

$$= 4.6\ (\$\ '0000)$$

Interpretation

Slope (B_1= 0.56) here means for an increase of one year in age of the person, the salary also increases by an amount of \$5600.

Intercept does not always make sense in terms of deriving meaning out of its value. Like in this example, the value of negative 6.6 suggests that if the person is not yet born (age = 0), the salary of that person would be negative \$66000.

Figure 4-9 shows the final regression line for our dataset.

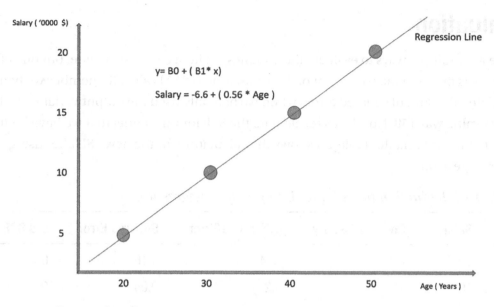

Figure 4-9. *Regression line*

Let's predict the salary for all four records in our data using the regression equation and compare these with actual salaries and see the difference as shown in Table 4-3.

Table 4-3. *Difference Between Predictions and Actual Values*

Age	Salary	Predicted Salary	Difference/Error
20	5	4.6	0.4
30	10	10.2	0.2
40	15	15.8	0.8
50	22	21.4	-0.6

In a nutshell, linear regression comes up with the most optimal values for the intercept (B_0) and coefficients (B_0, B_1, B_2) so that the difference (error) between the predicted values and the target variables is minimum.

But the question remains: Is it a good fit?

Evaluation

There are multiple ways to evaluate the goodness of fit of a regression line, but one of the ways is by using the coefficient of determination (r^{square}) value. Remember we had calculated the sum of squared errors when we had only used the output variable itself, and its value was 158. Now let us recalculate the SSE for this model that we have built using the input variable. Table 4-4 shows the calculation for the new SSE after using linear regression.

Table 4-4. *Reduction in SSE After Using Linear Regression*

Age	Salary	Predicted Salary	Difference/Error	Squared Error	Old SSE
20	5	4.6	-0.4	0.16	64
30	10	10.2	0.2	0.04	9
40	15	15.8	0.8	0.64	4
50	22	21.4	-0.6	0.36	81

As we can observe, the total sum of squared errors has reduced significantly from 158 to only 1.2, which has happened because of using linear regression. The variance in the target variable (salary) can be explained with the help of regression (due to usage of the input variable – age). So OLS works toward reducing the overall sum of squared errors. The total sum of squared errors is a combination of two types:

TSS (Total Sum of Squared Errors) = SSE (Sum of squared errors) + SSR (Residual sum of squared errors)

The TSS is the sum of the squared difference between the actual and the mean values, and it's always fixed. This was equal to 158 in our example.

The SSE is the squared difference from actual to predicted values of target variable, which reduced to 1.2 after using linear regression.

SSR is the sum of squared errors explained by regression and can be calculated by TSS – SSE:

$$SSR = 158 - 1.2 = 156.8$$

$$r^{square} \text{ (Coefficient of determination)} = \frac{SSR}{TSS} == \frac{156.8}{158} = 0.99$$

This percentage indicates that our linear regression model can predict with 99% accuracy in terms of predicting the salary amount given the age of the person. The other 1% can be attributed toward errors, which cannot be explained by the model. Our linear regression line fits the model really well, but it can also be a case of overfitting. Overfitting occurs when your model predicts with high accuracy on training data but its performance drops on unseen/test data. The technique to address the issue of overfitting is known as regularization, and there are different types of regularization techniques. In terms of linear regression, one can use the Ridge, Lasso, or Elastic Net regularization technique to handle overfitting.

Ridge regression is also known as L2 regularization and focuses on restricting the coefficient values of input features close to zero, whereas Lasso regression (L1) makes some of the coefficients to zero in order to improve generalization of the model. Elastic Net is the combination of both techniques.

At the end of the day, regression is a still a parameter-driven approach and assumes few underlying patterns about distributions of input data points. If the input data does not affiliate with those assumptions, the linear regression model does not perform well. Hence, it is important to go over these assumptions very quickly in order to know them before using the linear regression model:

- There must be a linear relationship between input variable and output variable.

- The independent variables (input features) should not be correlated to each other (also known as multicollinearity).

- There must be no correlation between the residuals/error values.

- There must be a linear relationship between the residuals and the output variable

- The residuals/error values must be normally distributed.

Code

This section of the chapter focuses on building a linear regression model from scratch using Pyspark and automating the steps using a pipeline. We saw a simple example of only one input variable to understand linear regression, but it is seldom the case. Majority of the times, the dataset would contain multiple variables, and hence building a multi-variable regression model makes more sense in such a situation. The linear regression equation then looks something like this:

$$y = B0 + B1 * X1 + B2 * X2 + B3 * X3 + \cdots$$

Note   The complete dataset along with the code is available for reference on the GitHub repo of this book and executes best on Spark 2.7 or higher.

Let's build a linear regression model using Spark's MLlib library and predict the target variable using the input features. The dataset that we are going to use for this example is a small dataset and contains a total of 1232 rows and 6 columns. We have to use five input variables to predict the target variable using the linear regression model.

We start the Jupyter notebook and import SparkSession and create a new SparkSession object to use Spark:

```
[In]: import pyspark
[In]: from pyspark.sql import SparkSession
[In]: spark=SparkSession.builder.appName("RegressionwithPySpark").
getOrCreate()
```

We can now read the dataset within Spark using the Dataframe. Since we're using Databricks, we can mention the file location of the dataset:

```
[In]: file_location = "/FileStore/tables/linear_reg_dataset.csv"
[In]: file_type = "csv"
[In]: infer_schema = "true"
[In]: first_row_is_header = "true"
[In]: delimiter = ","
[In]: df = spark.read.format(file_type) \
        .option("inferSchema", infer_schema) \
        .option("header", first_row_is_header) \
        .option("sep", delimiter) \
        .load(file_location)

[In]: display(df)
[Out]:
```

▸ ▦ df: pyspark.sql.dataframe.DataFrame = [var_1: integer, var_2: integer ... 4 more fields]

	var_1	var_2	var_3	var_4	var_5	output
1	734	688	81	0.328	0.259	0.418
2	700	600	94	0.32	0.247	0.389
3	712	705	93	0.311	0.247	0.417
4	734	806	69	0.315	0.26	0.415
5	613	759	61	0.302	0.24	0.378
6	748	676	85	0.318	0.255	0.422

In this section, we drill deeper into the dataset by viewing the dataset, validating the shape of the dataset, and discussing various statistical measures and correlations among input and output variables. We start with checking the shape of the dataset:

```
[In]: print((df.count(), len(df.columns)))
[Out]:
```

▸ (2) Spark Jobs

(1232, 6)

The preceding output confirms the size of our dataset, and we can validate the datatypes of the input values to check if we need to change/cast any column datatypes. In this example, all columns contain integer or double values:

```
[In]: df.printSchema()
[Out]:
```

```
root
 |-- var_1: integer (nullable = true)
 |-- var_2: integer (nullable = true)
 |-- var_3: integer (nullable = true)
 |-- var_4: double (nullable = true)
 |-- var_5: double (nullable = true)
 |-- output: double (nullable = true)
```

There are a total of six columns, out of which five are input columns (var_1 to var_5), with one target column (output). We can now use the describe function to go over statistical measures of the dataset:

```
[In]: df.describe().show(3,False)
[Out]:
```

summary	var_1	var_2	var_3	var_4	var_5	output
count	1232	1232	1232	1232	1232	1232
mean	715.0819805194806	715.0819805194806	80.90422077922078	0.3263311688311693	0.25927272727272715	0.39734172077922014
stddev	91.5342940441652	93.07993263118064	11.458139049993724	0.015012772334166148	0.012907228928000298	0.03326689862173776

This allows us to get a sense of distribution and the measure of center and spread for our dataset columns. For keeping things simple, let's rename the output column to label as it becomes relatively easier to compare the model performance on test data:

```
[In]: df=df.withColumnRenamed('output','label')
[In]: df.show()
[Out]:
```

```
+-----+-----+-----+-----+-----+-----+
|var_1|var_2|var_3|var_4|var_5|label|
+-----+-----+-----+-----+-----+-----+
|  734|  688|   81|0.328|0.259|0.418|
|  700|  600|   94| 0.32|0.247|0.389|
|  712|  705|   93|0.311|0.247|0.417|
|  734|  806|   69|0.315| 0.26|0.415|
|  613|  759|   61|0.302| 0.24|0.378|
|  748|  676|   85|0.318|0.255|0.422|
|  669|  588|   97|0.315|0.251|0.411|
|  667|  845|   68|0.324|0.251|0.381|
|  758|  890|   64| 0.33|0.274|0.436|
|  726|  670|   88|0.335|0.268|0.422|
|  583|  794|   55|0.302|0.236|0.371|
|  676|  746|   72|0.317|0.265|  0.4|
|  767|  699|   89|0.332|0.274|0.433|
|  637|  597|   86|0.317|0.252|0.374|
|  609|  724|   69|0.308|0.244|0.382|
|  776|  733|   83|0.325|0.259|0.437|
|  701|  832|   66|0.325| 0.26| 0.39|
```

We can check correlation between input variables and output variable using the corr function. In the following example, we validate the correlation between var_1 and label. As we can see, it is a pretty strong correlation:

```
[In]: from pyspark.sql.functions import corr
[In]: df.select(corr('var_1','label')).show()
[Out]:
```

```
+------------------+
|corr(var_1, label)|
+------------------+
|0.9187399607627283|
+------------------+
```

```
[In]: from pyspark.ml.linalg import Vector
[In]: from pyspark.ml.feature import VectorAssembler
```

This is the part where we create a single vector combining all input features by using Spark's VectorAssembler. It creates only a single feature that captures the input values for that row. So, instead of five input columns, it essentially merges all input columns into a single feature vector column. One can select the number of columns that would be used as input features and can pass only those columns through VectorAssembler. In our case, we will pass all the five input columns to create a single feature vector column:

```
[In]:vec_assmebler=VectorAssembler(inputCols=['var_1','var_2','var_3','var_
4','var_5'],outputCol='features')
[In]: features_df=vec_assmebler.transform(df)
[In]: features_df.printSchema()
[Out]:
```

```
root
 |-- var_1: integer (nullable = true)
 |-- var_2: integer (nullable = true)
 |-- var_3: integer (nullable = true)
 |-- var_4: double (nullable = true)
 |-- var_5: double (nullable = true)
 |-- label: double (nullable = true)
 |-- features: vector (nullable = true)
```

As we can see, we have an additional column ("features"), which contains the single dense vector for all of the inputs:

```
[In]: features_df.select('features').show(5,False)
[Out]:
```

```
+------------------------------+
|features                      |
+------------------------------+
|[734.0,688.0,81.0,0.328,0.259]|
|[700.0,600.0,94.0,0.32,0.247] |
|[712.0,705.0,93.0,0.311,0.247]|
|[734.0,806.0,69.0,0.315,0.26] |
|[613.0,759.0,61.0,0.302,0.24] |
+------------------------------+
```

We take a subset of the Dataframe and select only the features column and the output column to build the linear regression model:

```
[In]: model_df=features_df.select('features','label')
[In]: model_df.show(5)
[Out]:
```

```
+--------------------+-----+
|            features|label|
+--------------------+-----+
|[734.0,688.0,81.0...|0.418|
|[700.0,600.0,94.0...|0.389|
|[712.0,705.0,93.0...|0.417|
|[734.0,806.0,69.0...|0.415|
|[613.0,759.0,61.0...|0.378|
+--------------------+-----+
```

```
[In]: print((model_df.count(), len(model_df.columns)))
[Out]: (1232, 2)
```

We have to split the dataset into training and test data in order to train and evaluate the performance of the linear regression model built. We split it in 75/25 ratio and train our model on 70% of the dataset. We can print the shape of train and test data to validate the size:

```
[In]: train_df,test_df=model_df.randomSplit([0.75,0.25])
[In]: print((train_df.count(), len(train_df.columns)))
[Out]: (915,2)
[In]: print((test_df.count(), len(test_df.columns)))
[Out]: (317,2)
```

In this part, we build and train the linear regression model using the features and output columns. We can fetch the coefficient (B1, B2, B3, B4, B5) and intercept (B0) values of the model as well. We can also evaluate the performance of the model on training and test data using r2 and RMSE:

```
[In]: from pyspark.ml.regression import LinearRegression
[In]: lin_Reg=LinearRegression(labelCol='label')
[In]: lr_model=lin_Reg.fit(train_df)
[In]: print(lr_model.coefficients)
[Out]: [0.00034145505732588027,5.048358349575795e-05,
0.00019743563636407933,-0.6583737151894117,0.4837227208356984]
[In]: print(lr_model.intercept)
[Out]: 0.1905629807694056
[In]: training_predictions=lr_model.evaluate(train_df)
[In]: print(training_predictions.r2)
[Out]: 0.8621803970011387
```

As we can see, the model is doing a decent job of predicting on the training data with r2 value of around 86%. The final part of the modeling exercise is to check the performance of the model on unseen or test data. We use the evaluate function to make predictions for test data and can use r2 and RMSE to check the accuracy of the model on test data. The performance seems to get bit better than that of training:

```
[In]: test_results=lr_model.evaluate(test_df)
[In]: print(test_results.r2)
[Out]: 0.8873805060206182
[In]: print(test_results.rootMeanSquaredError)
[Out]: 0.011586329962905937
```

Now that we have seen the steps to run the regression model, we can make use of a pipeline to automate some of these steps. We can leverage the power of Spark pipelines to do all the preprocessing, feature engineering, and model training for us. We only need to ensure that we declare the steps that need to be followed in a particular sequence.

We start by creating a fresh Spark Dataframe using the same source data. We call it as new_df this time around. We then rename the target column to label as done before. The final step before we move on to the pipeline is to split the data into train and test – we use the same ratio as used before:

```
[In]: new_df = spark.read.format(file_type) \
        .option("inferSchema", infer_schema) \
        .option("header", first_row_is_header) \
        .option("sep", delimiter) \
        .load(file_location)

[In]: new_df=new_df.withColumnRenamed('output','label')
[In]: train_df, test_df = new_df.randomSplit([.75, .25])
```

Now we can start declaring the different stages of the pipeline for end-to-end prediction. We need to import Pipeline from Pyspark. Just to demonstrate the strength of pipelines, let us add scaling as an additional data processing step:

```
[In]: from pyspark.ml.feature import StandardScaler
[In]: from pyspark.ml import Pipeline
```

In essence, now we declare the different stages for our pipeline. We start with declaring the first stage where we create the feature vector using VectorAssembler followed by standardizing the input values using StandardScaler. We then apply the desired algorithm (linear regression in this case) on the training data. The pipeline can have any number of stages depending on the workflow:

```
[In]: features=['var_1', 'var_2', 'var_3', 'var_4', 'var_5']
[In]: stage_1 = VectorAssembler(inputCols=features, outputCol="out_
features")
[In]: stage_2 = StandardScaler(inputCol="out_features",
outputCol="features")
[In]: stage_3 = LinearRegression()
[In]: stages = [stage_1, stage_2, stage_3]
[In]: pipeline = Pipeline(stages=stages)
```

Now that we have defined the pipeline stages, we can fit it on the training data and transform on test data. We end up with a Dataframe that contains the model prediction column. It also contains the feature vector:

```
[In]: model = pipeline.fit(train_df)
[In]: pred_result= model.transform(test_df)
[In]: pred_result.show(5)
[Out]:
```

```
+-----+-----+-----+-----+-----+-----+--------------------+--------------------+-------------------+
|var_1|var_2|var_3|var_4|var_5|label|        out_features|            features|         prediction|
+-----+-----+-----+-----+-----+-----+--------------------+--------------------+-------------------+
|  495|  628|   66|0.285|0.229|0.315|[495.0,628.0,66.0...|[5.47518903568311...|0.32718990066573794|
|  495|  752|   50|0.277|0.221|0.327|[495.0,752.0,50.0...|[5.47518903568311...| 0.3328023411526273|
|  498|  615|   67|0.291|0.227|0.318|[498.0,615.0,67.0...|[5.50837199953574...| 0.3228284746165683|
|  511|  576|   76| 0.29|0.231|0.329|[511.0,576.0,76.0...|[5.65216484289711...|0.32916764726139747|
|  532|  690|   69|0.303|0.245|0.351|[532.0,690.0,69.0...|[5.88444558986549...| 0.3395640235921174|
+-----+-----+-----+-----+-----+-----+--------------------+--------------------+-------------------+
```

Now that we have the predictions on the test data, we can evaluate the accuracy of the linear regression model using r2 and RMSE:

```
[In]: from pyspark.ml.evaluation import RegressionEvaluator
[In]: regeval = RegressionEvaluator(labelCol="label",
predictionCol="prediction", metricName="rmse")
```

```
[In]: acc = regeval.evaluate(pred_result, {regeval.metricName: "r2"})
[In]: print(acc)
[Out]: 0.8705195703144362
[In]: rmse = regeval.evaluate(pred_result)
[In]: print(rmse)
[Out]: 0.01208473164101912
```

The model seems to be pretty consistent with the overall accuracy of 87% of r2. The slight variation that we observe in the r2 value here is due to the fact that data is split on a random basis. So using Spark pipelines, we were able to automate bulk of the steps involved in the model's end-to-end prediction. In the coming chapters, we further build on this to see how pipelines can help reduce chances of errors and make the overall prediction process seamless.

Conclusion

In this chapter, we went over the fundamentals of linear regression along with the approach of building a regression model in PySpark. We also covered the process of automating the steps for end-to-end predictions.

CHAPTER 5

Logistic Regression

This chapter focuses on building a logistic regression model with Pyspark along with understanding the ideas behind logistic regression. Logistic regression is used for classification problems. We have already seen classification details in earlier chapters. Although it is used for classification, still it's called logistic regression. It is due to the fact that under the hood, linear regression equations still operate to find the relationship between input variables and target variables. The main distinction between linear and logistic regression is that we use some sort of nonlinear function to convert the output of the latter into a probability to restrict it between 0 and 1. For example, we can use logistic regression to predict if a user would buy the product or not. In this case, the model would return a buying probability for each user. Logistic regression is used widely in many business applications.

Probability

To understand logistic regression, we will have to go over the concept of probability first. It is defined as the chances of occurrence of a desired event or interested outcomes upon all possible outcomes. Take for an example if we flip a coin. The chances of getting heads or tails are equal (50%) as shown in Figure 5-1.

If we roll a fair dice, then the probability of getting any of the number between 1 and 6 is equal to 16.7%.

If we pick a ball from a bag that contains four green balls and one blue ball, the probability of picking a green ball is 80%.

© Pramod Singh 2022
P. Singh, *Machine Learning with PySpark*, https://doi.org/10.1007/978-1-4842-7777-5_5

Figure 5-1. *Probability of events*

Logistic regression is used to predict the probability of each target class. In the case of binary classification (only two classes), it returns the probability associated with each class for every record. As mentioned, it uses linear regression behind the scenes in order to capture the relationship between input and output variables, yet we additionally use one more element (nonlinear function) to convert the output from a continuous form into probability. Let's understand this with the help of an example. Let's consider that we have to use models to predict if some particular user would buy the product or not and we are using only a single input variable that is time spent by the user on the website. The data for the same is given in Table 5-1.

Table 5-1. *Sample Data*

Sr. No	Time Spent (mins)	Converted
1	1	No
2	2	No
3	5	No
4	15	Yes
5	17	Yes
6	18	Yes

Let us visualize this data in order to see the distinction between converted and non-converted users as shown in Figure 5-2.

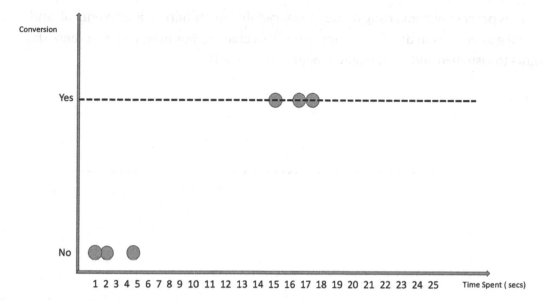

Figure 5-2. *Conversion status vs. time spent*

Using Linear Regression

Let's try using linear regression instead of logistic regression to understand the reasons why logistic regression makes more sense in classification scenarios. In order to use linear regression, we will have to convert the target variable from categorical into numeric form. So let's reassign the values for the Converted column:

Yes = 1
No = 0

Now, our data looks something as given in Table 5-2.

Table 5-2. *Regression Output*

Sr. No	Time Spent (mins)	Converted
1	1	0
2	2	0
3	5	0
4	15	1
5	17	1
6	18	1

This process of converting a categorical variable to numerical is also critical, and we will go over this in detail in a later part of this chapter. For now, let's plot these data points to visualize and understand it better (Figure 5-3).

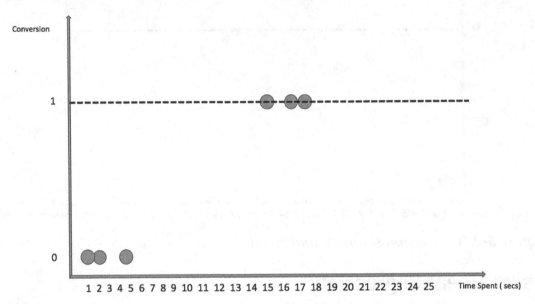

Figure 5-3. *Conversion status (1 and 0) vs. time spent*

As we can observe, there are only two values in our target column (1 and 0), and every point lies on either of these two values. Now, let's suppose we do linear regression on these data points and come up with a "best-fit line," which is shown in Figure 5-4.

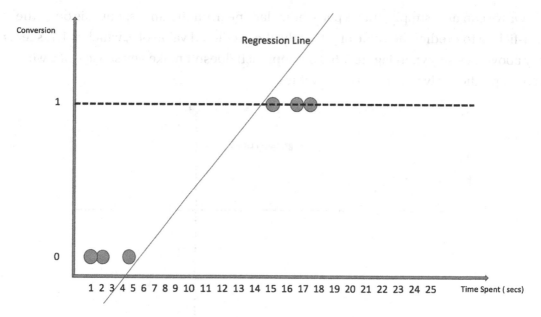

Figure 5-4. *Regression line for users*

The regression equation for this line would be

$$y = B_0 + B_1 * x$$

$$y_{(1,0)} = B_0 + B_1 * Time\ Spent$$

All looks good so far in terms of coming up with a straight line to distinguish between 1 and 0 values. It seems like linear regression is also doing a good job of differentiating between converted and non-converted users, but there is a slight problem with this approach.

Take for an example a new user spends 20 seconds on the website and we have to predict if this user will convert or not using the linear regression line. We use the preceding regression equation and try to predict the y value for 20 seconds of time spent.

We can simply calculate the value of y by either calculating

$$y = B_0 + B_1 * (20)$$

Or we can also simply draw a perpendicular line from the time spent axis on to the best-fit line to predict the value of y. Clearly, the predicted value of y, which is 1.7, seems way above 1 as shown in Figure 5-5. This approach doesn't make any sense since we want to predict only between 0 and 1 values.

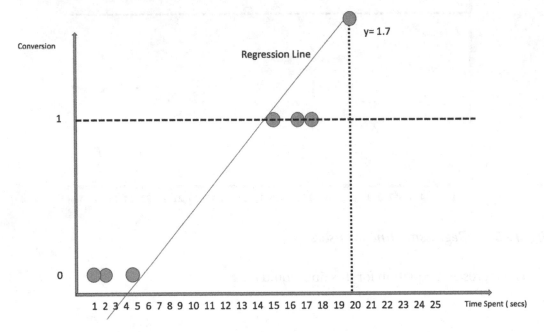

Figure 5-5. *Predictions using a regression line*

So, if we use linear regression for classification cases, it creates a situation where the predicted output values can range from –infinity to +infinity. Hence, we need another approach that can tie these values between 0 and 1 only. The notion of values between 0 and 1 is not unfamiliar anymore as we have already seen probability. So, essentially, logistic regression comes up with a decision boundary between positive and negative classes that is associated with a probability value.

Using Logit

To accomplish the objective of converting the output value into probability, we use something called Logit. Logit is a nonlinear function and does a nonlinear transformation of a linear equation to convert the output between 0 and 1. In logistic regression, that nonlinear function is a sigmoid function, which looks like this:

$$\frac{1}{1+e^{-x}}$$

And it always produces values between 0 and 1 independent of values of x.
So let's go back to our earlier linear regression equation

$$y = B_0 + B_1 * Time\ Spent$$

We pass our output (y) through this nonlinear function (sigmoid) to change its values between 0 and 1:

$$Probability = \frac{1}{1 + e^{-y}}$$

$$Probability = \frac{1}{1 + e^{-(B_0 + B_1 * Time\ Spent)}}$$

Using the preceding equation, the predicted value gets limited between 0 and 1, and the output now looks as shown in Figure 5-6.

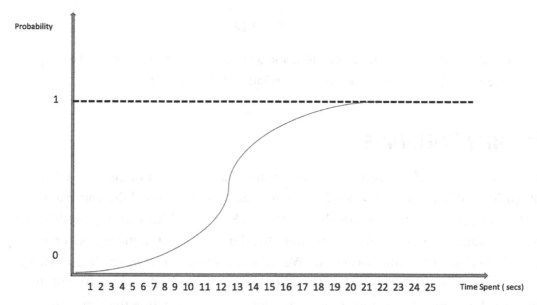

Figure 5-6. *Logistic curve*

The advantage of using the nonlinear function is that irrespective of any value of input (time spent), the output would always be the probability of conversion. This curve is also known as logistic curve. Logistic regression also assumes that there is a linear relationship between the input and the target variables, and hence the most optimal values of the intercept and coefficients are found out to capture this relationship.

Interpretation (Coefficients)

The coefficients of the input variables are found using a technique known as gradient descent, which looks for optimizing the loss function in such a way that the total error is minimized. We can look at the logistic regression equation and understand the interpretation of coefficients:

$$y = \frac{1}{1 + e^{-(B_0 + B_1 * x)}}$$

Let's say after calculating for the data points in our example, we get the coefficient value of time spent as 0.75.

In order to understand what this 0.75 means, we have to take the exponential value of this coefficient:

$$e^{0.75} = 2.12$$

This 2.12 is known as odd ratio, and it suggests that per unit increase in time spent on the website, the odds of customer conversion increase by 112%.

Dummy Variables

So far, we have only dealt with continuous/numerical variables, but the presence of categorical variables in the dataset is inevitable. So let's understand the approach to use the categorical values for modeling purposes. Since machine learning models only consume data in numerical format, we have to adopt some technique to convert the categorical data into a numerical form. We have already seen one example previously where we converted our target class (Yes/No) into numerical values (1 or 0). This is known as label encoding where we assign unique numerical values to each of the category present in that particular column. There is another approach that works really well known as dummification or one-hot encoding. Let's understand this with the help of an example. Let's add one more column to our existing example data. Suppose we have one additional column that contains the search engine the user used. So our data looks something as shown in Table 5-3.

Table 5-3. *Additional Data Column*

Sr. No	Time Spent (mins)	Search Engine	Converted
1	5	Google	0
2	2	Bing	0
3	10	Yahoo	1
4	15	Bing	1
5	1	Yahoo	0
6	12	Google	1

So, to consume the additional information provided in the Search Engine column, we have to convert this into numerical format using dummification. As a result, we would get an additional number of dummy variables (columns), which would be equal to the number of distinct categories in the Search Engine column. The following steps explain the entire process of converting a categorical feature into numerical:

1. Find out the distinct number of categories in a categorical column. We have only three distinct categories as of now (Google, Bing, Yahoo).

2. Create new columns for each of the distinct categories and add value 1 in the category column for when the corresponding search engine is used or else 0 as shown in Table 5-4.

Table 5-4. *Column Representation*

Sr. No	Time Spent (mins)	Search Engine	SE_Google	SE_Bing	SE_Yahoo	Converted
1	1	Google	1	0	0	0
2	2	Bing	0	1	0	0
3	5	Yahoo	0	0	1	0
4	15	Bing	0	1	0	1
5	17	Yahoo	0	1	0	1
6	18	Google	1	0	0	1

3. Remove the original category column. So the dataset now contains five columns in total (excluding index) because we have three additional dummy variables as shown in Table 5-5.

Table 5-5. *Refined Column Representation*

Sr. No	Time Spent (mins)	SE_Google	SE_Bing	SE_Yahoo	Converted
1	1	1	0	0	0
2	2	0	1	0	0
3	5	0	0	1	0
4	15	0	1	0	1
5	17	0	1	0	1
6	18	1	0	0	1

The whole idea is to represent the same information in a different manner so that the machine learning model can learn from categorical values as well.

Model Evaluation

To measure the performance of the logistic regression model, we can use multiple metrics. The most obvious one is the accuracy parameter. Accuracy is the percentage of correct predictions made by the model. However, accuracy is not always the preferred approach. To understand the performance of the logistic model, we should use a confusion matrix. It consists of the value counts for the predictions vs. actual values. A confusion matrix for a binary class looks like Table 5-6.

Table 5-6. *Confusion Matrix*

Actual/Prediction	Predicted Class (Yes)	Predicted Class (No)
Actual Class (Yes)	True Positives (TP)	False Negatives (FN)
Actual Class (No)	False Positives (FP)	True Negatives (TN)

Let us understand the individual values in the confusion matrix.

True Positives

These are the values that are of the positive class in actuality, and the model also correctly predicted them to be of the positive class.

- Actual Class: Positive (1)
- ML Model Prediction Class: Positive (1)

True Negatives

These are the values that are of negative class in actuality, and the model also correctly predicted them to be of the negative class.

- Actual Class: Negative (0)
- ML Model Prediction Class: Negative (1)

False Positives

These are the values that are of the negative class in actuality, but the model incorrectly predicted them to be of the positive class.

- Actual Class: Negative (0)
- ML Model Prediction Class: Positive (1)

False Negatives

These are the values that are of the positive class in actuality, but the model incorrectly predicted them to be of the negative class.

- Actual Class: Positive (1)
- ML Model Prediction Class: Negative (1)

Accuracy

Accuracy is the sum of true positives and true negatives divided by the total number of records:

$$\frac{(TP+TN)}{Total\ number\ of\ Records}$$

But as said earlier, it is not always the preferred metric because of the target class imbalance. Most of the times, target class frequency is skewed (more number of TN examples compared to TP examples). Take for an example the dataset for fraud detection contains 99% of genuine transactions and only 1% fraud ones. Now, if our logistic regression model predicts all genuine transactions and no fraud case, it still ends up with 99% accuracy. The whole point is to find out the performance in regard to the positive class. Hence, there are a couple of other evaluation metrics that we can use.

Recall

Recall rate helps in evaluating the performance of the model from a positive class standpoint. It tells the percentage of actual positive cases the model is able to predict correctly out of the total number of positives cases:

$$\frac{(TP)}{(TP+FN)}$$

It talks about the quality of the machine learning model when it comes to predicting the positive class. So out of the total positive class, it tells how many the model was able to predict correctly. This metric is widely used as an evaluation criterion for classification models.

Precision

Precision is about the number of actual positives cases out of all the positive cases predicted by the model:

$$\frac{(TP)}{(TP+FP)}$$

This can also be used as an evaluation criterion.

F1 Score

$$\text{F1 Score} - 2 * \frac{(Precision * Recall)}{(Precision + Recall)}$$

Probability Cut-Off/Threshold

Since we know the output of the logistic regression model is a probability score, it is very important to decide the cut-off or threshold limit of probability for prediction. By default, the probability threshold is set at 50%. It means, if the probability output of the model is below 50%, the model will predict it to be of the negative class (0) and, if it is equal and above 50%, it would be assigned the positive class (1).

If the threshold limit is very low, then the model will predict a lot of positive classes and would have a high recall rate. On the contrary, if the threshold limit is very high, then the model might miss out on positive cases, and the recall rate would be low, but precision would be higher. In this case, the model will predict very few positive cases. Deciding a good threshold value is often challenging. A Receiver Operator Characteristic curve, or ROC curve, can help to decide which value of the threshold is best.

ROC Curve

The ROC is used to decide the threshold value for the model. It is the plot between recall (also known as sensitivity) and precision (specificity) as shown in Figure 5-7.

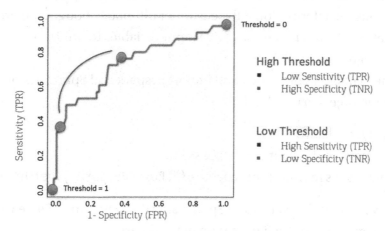

Figure 5-7. *ROC curve*

One would like to pick a threshold that offers a balance between both recall and precision. So, now that we have understood various components associated with logistic regression, we can go ahead and build a logistic regression model using PySpark.

Logistic Regression Code

This section of the chapter focuses on building a logistic regression model from scratch using PySpark and a Jupyter notebook.

Note   The complete dataset along with the code is available for reference on the GitHub repo of this book and executes best on Spark 3.1 or higher.

Let's build a logistic regression model using Spark's MLlib library and predict the target class label.

Data Info

The dataset that we are going to use is a sample dataset that contains a total of 20000 rows and 6 columns. This dataset contains information regarding online users of a retail sports merchandise company. The data captures the country of the user, platform used, age, repeat visitor or first-time visitor, and number of web pages viewed at the website. It also has the information if the customer ultimately bought the product or not (conversion status). We will make use of five input variables to predict the target class using a logistic regression model.

We start the Databricks notebook and import pyspark and SparkSession to create a new SparkSession object to use Spark:

```
[In]: import pyspark
[In]: from pyspark.sql import SparkSession
[In]:spark=SparkSession.builder.appName("LRwithPySpark").getOrCreate()
```

We can now read the dataset within Spark using the Dataframe. Since we're using Databricks, we can mention the file location of the dataset:

```
[In]: file_location = "/FileStore/tables/data.csv"
[In]: file_type = "csv"
```

```
[In]: infer_schema = "true"
[In]: first_row_is_header = "true"
[In]: delimiter = ","
[In]: df = spark.read.format(file_type) \
        .option("inferSchema", infer_schema) \
        .option("header", first_row_is_header) \
        .option("sep", delimiter) \
        .load(file_location)

[In]: display(df)
[Out]:
```

	Country	Age	Repeat_Visitor	Platform	Web_pages_viewed	Status
1	India	41	1	Yahoo	21	1
2	Brazil	28	1	Yahoo	5	0
3	Brazil	40	0	Google	3	0
4	Indonesia	31	1	Bing	15	1
5	Malaysia	32	0	Google	15	1
6	Brazil	32	0	Google	3	0

Now we look deeper into the dataset by viewing the dataset, validating the shape of the dataset, and discussing various statistical measures of the variables. We start with checking the shape of the dataset:

```
[In]:print((df.count(), len(df.columns)))
[Out]: (20000, 6)
```

The output confirms the size of our dataset. Now we can then validate the datatypes of the input values to check if we need to change/cast any attribute's datatype:

```
[In]: df.printSchema()
        root
         |-- Country: string (nullable = true)
         |-- Age: integer (nullable = true)
         |-- Repeat_Visitor: integer (nullable = true)
         |-- Platform: string (nullable = true)
         |-- Web_pages_viewed: integer (nullable = true)
         |-- Status: integer (nullable = true)
```

As we can see, there are two columns (Country, Platform) that are categorical in nature and hence need to be converted into numerical form later. We can now use the describe function to go over statistical measures of the dataset:

```
[In]: df.describe().show()
[Out]:
```

```
+-------+-------+------------------+------------------+--------+------------------+------------------+
|summary|Country|Age               |Repeat_Visitor    |Platform|Web_pages_viewed  |Status            |
+-------+-------+------------------+------------------+--------+------------------+------------------+
|count  |20000  |20000             |20000             |20000   |20000             |20000             |
|mean   |null   |28.53955          |0.5029            |null    |9.5533            |0.5               |
|stddev |null   |7.888912950773227 |0.500004090187782 |null    |6.073903499824976 |0.5000125004687693|
+-------+-------+------------------+------------------+--------+------------------+------------------+
```

As we can observe, the average age of visitors is close to 28 years, and they view around nine web pages during the website visit. Let us explore individual columns to understand data in more detail. The groupBy function used along with count returns us the frequency of each of the categories in the data. This is similar to value_counts in Pandas:

```
[In]: df.groupBy('Country').count().show()
[Out]:
```

```
+---------+-----+
|  Country|count|
+---------+-----+
| Malaysia| 1218|
|    India| 4018|
|Indonesia|12178|
|   Brazil| 2586|
+---------+-----+
```

So the maximum number of visitors are from Indonesia followed by India.

```
[In]: df.groupBy('Platform').count().show()
[Out]:
```

```
+--------+-----+
|Platform|count|
+--------+-----+
|   Yahoo| 9859|
|    Bing| 4360|
|  Google| 5781|
+--------+-----+
```

The Yahoo platform users are the highest in number compared with the other platforms.

```
[In]: df.groupBy('Status').count().show()
[Out]:
+------+-----+
|Status|count|
+------+-----+
|     1|10000|
|     0|10000|
+------+-----+
```

We seem to have a balanced target class in this dataset as there are an equal number of users who have converted and not converted. Let's use the groupBy function along with mean to know more about the dataset:

```
[In]: df.groupBy('Country').mean().show()
[Out]:
```

Country	avg(Age)	avg(Repeat_Visitor)	avg(Web_pages_viewed)	avg(Status)
Malaysia	27.792282430213465	0.5730706075533661	11.192118226600986	0.6568144499178982
India	27.976854156296664	0.5433051269288203	10.727227476356397	0.6212045793927327
Indonesia	28.43159796354081	0.5207751683363442	9.985711939563148	0.5422893742814913
Brazil	30.274168600154677	0.322892498066512	4.921113689095128	0.038669760247486466

We have the highest conversion rate from Malaysia followed by India. The average number of web page visits is highest in Malaysia and lowest in Brazil.

```
[In]: df.groupBy('Platform').mean().show()
[Out]:
```

Platform	avg(Age)	avg(Repeat_Visitor)	avg(Web_pages_viewed)	avg(Status)
Yahoo	28.569226087838523	0.5094837204584644	9.599655137437875	0.5071508266558474
Bing	28.68394495412844	0.4720183486238532	9.114908256880733	0.4559633027522936
Google	28.380038055699707	0.5149628092025601	9.804878048780488	0.5210171250648676

We have the highest conversion rate from users of the Google platform followed by Yahoo.

```
[In]: df.groupBy('Status').mean().show()
[Out]:
```

```
+------+--------+------------------+------------------+-----------+
|Status|avg(Age)|avg(Repeat_Visitor)|avg(Web_pages_viewed)|avg(Status)|
+------+--------+------------------+------------------+-----------+
|     1| 26.5435|            0.7019|           14.5617|        1.0|
|     0| 30.5356|            0.3039|            4.5449|        0.0|
+------+--------+------------------+------------------+-----------+
```

We can see there is a strong connection between the conversion status and number of pages viewed along with repeat visits.

Now we move on to convert the categorical variables into numerical form using an encoder. We then use VectorAssembler to create a single vector combining all input features:

```
[In]: from pyspark.ml.feature import StringIndexer
[In]: from pyspark.ml.feature import VectorAssembler
```

Since we are dealing with two categorical columns, we will have to convert the Country and Platform columns into numerical form.

The first step is to label the column using StringIndexer into numerical form. It allocates unique values to each of the categories of the column. So, in the following example, all the three values of Platform (Yahoo, Google, Bing) are assigned values 0.0, 1.0, and 2.0. This is visible in the column named Platform_index. Similarly for the Country column, we can observe similar values in the Country_index column. The way it assigns values to unique categories is based on the value counts for each category. So, in the case of the Country column, Indonesia has the highest occurrences and hence gets assigned 0.0, followed by India with 1.0:

```
[In]: si_platform = StringIndexer(inputCol='Country',outputCol=
'Country_Index')
[In]: df = si_platform.fit(df).transform(df)
[In]: si_country =StringIndexer(inputCol='Platform',outputCol=
'Platform_Index')
```

```
[In]: df = si_country.fit(df).transform(df)
[In] df.show(10)
```

```
+---------+---+--------------+--------+----------------+------+-------------+--------------+
| Country|Age|Repeat_Visitor|Platform|Web_pages_viewed|Status|Country_Index|Platform_Index|
+---------+---+--------------+--------+----------------+------+-------------+--------------+
|    India| 41|             1|   Yahoo|              21|     1|          1.0|           0.0|
|   Brazil| 28|             1|   Yahoo|               5|     0|          2.0|           0.0|
|   Brazil| 40|             0|  Google|               3|     0|          2.0|           1.0|
|Indonesia| 31|             1|    Bing|              15|     1|          0.0|           2.0|
| Malaysia| 32|             0|  Google|              15|     1|          3.0|           1.0|
|   Brazil| 32|             0|  Google|               3|     0|          2.0|           1.0|
|   Brazil| 32|             0|  Google|               6|     0|          2.0|           1.0|
|Indonesia| 27|             0|  Google|               9|     0|          0.0|           1.0|
|Indonesia| 32|             0|   Yahoo|               2|     0|          0.0|           0.0|
|Indonesia| 31|             1|    Bing|              16|     1|          0.0|           2.0|
+---------+---+--------------+--------+----------------+------+-------------+--------------+
```

The next step is to represent each of these values in one form with a one-hot encoded vector. This new vector looks a little different compared with the traditional one-hot encoder in Pandas in terms of representation as it captures the values and positions of the values in the vector:

```
[In]: from pyspark.ml.feature import OneHotEncoder
```

```
[In]: encoder = OneHotEncoder(inputCols=['Country_Index', 'Platform_
Index'],outputCols=['Country_vec', 'Platform_vec'])
```

```
[In]: df = encoder.fit(df).transform(df)
```

```
[In]: df.show(10)
[Out]:
```

```
+---------+---+--------------+--------+----------------+------+-------------+--------------+--------------+--------------+
| Country|Age|Repeat_Visitor|Platform|Web_pages_viewed|Status|Country_Index|Platform_Index| Country_vec| Platform_vec|
+---------+---+--------------+--------+----------------+------+-------------+--------------+--------------+--------------+
|    India| 41|             1|   Yahoo|              21|     1|          1.0|0.0|(3,[1],[1.0])|(2,[0],[1.0])|
|   Brazil| 28|             1|   Yahoo|               5|     0|          2.0|0.0|(3,[2],[1.0])|(2,[0],[1.0])|
|   Brazil| 40|             0|  Google|               3|     0|          2.0|1.0|(3,[2],[1.0])|(2,[1],[1.0])|
|Indonesia| 31|             1|    Bing|              15|     1|          0.0|2.0|(3,[0],[1.0])|    (2,[],[])|
| Malaysia| 32|             0|  Google|              15|     1|          3.0|1.0|    (3,[],[])|(2,[1],[1.0])|
|   Brazil| 32|             0|  Google|               3|     0|          2.0|1.0|(3,[2],[1.0])|(2,[1],[1.0])|
|   Brazil| 32|             0|  Google|               6|     0|          2.0|1.0|(3,[2],[1.0])|(2,[1],[1.0])|
|Indonesia| 27|             0|  Google|               9|     0|          0.0|1.0|(3,[0],[1.0])|(2,[1],[1.0])|
|Indonesia| 32|             0|   Yahoo|               2|     0|          0.0|0.0|(3,[0],[1.0])|(2,[0],[1.0])|
|Indonesia| 31|             1|    Bing|              16|     1|          0.0|2.0|(3,[0],[1.0])|    (2,[],[])|
+---------+---+--------------+--------+----------------+------+-------------+--------------+--------------+--------------+
```

```
[In]:df.groupBy('Country_vec').count().orderBy('count',ascending=False).
show(5,False)
[Out]:
```

```
+--------------+-----+
|Country_vec   |count|
+--------------+-----+
|(3,[0],[1.0])|12178|
|(3,[1],[1.0])|4018 |
|(3,[2],[1.0])|2586 |
|(3,[],[])     |1218 |
+--------------+-----+
```

As we can observe, the count values are same for each category in the Country column before one-hot encoding. Let's interpret the new one-hot encoded vector to understand the components better.

(3,[0],[1.0]) represents a vector of length 3, with 1 value :

Size of vector: 3

Value contained in vector: 1.0

Position of 1.0 value in vector: 0th place

This kind of representation allows to save computational space and hence is faster to compute. The length of the vector is equal to one less than the total number of elements since each value can be easily represented with just the help of three columns:

```
[In]:df.groupBy('Platform_vec').count().orderBy('count',ascending=False).
show(5,False)
[Out]:
```

```
+--------------+-----+
|Platform_vec  |count|
+--------------+-----+
|(2,[0],[1.0])|9859 |
|(2,[1],[1.0])|5781 |
|(2,[],[])     |4360 |
+--------------+-----+
```

In the case of the Platform vector, we observe we just need a vector of size 2 as the total number of unique values in the Platform column are just three. Now that we have converted both the categorical columns into numerical forms, we need to assemble all of the input columns into one single vector that would act as the input feature for the model. So we select the input columns that we need to use to create the single feature vector and name the output vector as features:

```
[In]: from pyspark.ml.feature import VectorAssembler

[In]: df_assembler = VectorAssembler(inputCols=[ 'Age', 'Repeat_
Visitor','Web_pages_viewed','Country_vec','Platform_vec'],
outputCol="features")

[In}:df = df_assembler.transform(df)

[In]: df.show()
[Out]:
```

Country	Age	Repeat_Visitor	Platform	Web_pages_viewed	Status	Country_Index	Platform_Index	Country_vec	Platform_vec	features
India	41	1	Yahoo	21	1	1.0	0.0	(3,[1],[1.0])	(2,[0],[1.0])	[41.0,1.0,21.0,0....
Brazil	28	1	Yahoo	5	0	2.0	0.0	(3,[2],[1.0])	(2,[0],[1.0])	[28.0,1.0,5.0,0.0...
Brazil	40	0	Google	3	0	2.0	1.0	(3,[2],[1.0])	(2,[1],[1.0])	(8,[0,2,5,7],[40....
Indonesia	31	1	Bing	15	1	0.0	2.0	(3,[0],[1.0])	(2,[],[])	(8,[0,1,2,3],[31....
Malaysia	32	0	Google	15	1	3.0	1.0	(3,[],[])	(2,[1],[1.0])	(8,[0,2,7],[32.0,...
Brazil	32	0	Google	3	0	2.0	1.0	(3,[2],[1.0])	(2,[1],[1.0])	(8,[0,2,5,7],[32....
Brazil	32	0	Google	6	0	2.0	1.0	(3,[2],[1.0])	(2,[1],[1.0])	(8,[0,2,5,7],[32....
Indonesia	27	0	Google	9	0	0.0	1.0	(3,[0],[1.0])	(2,[1],[1.0])	(8,[0,2,3,7],[27....
Indonesia	32	0	Yahoo	2	0	0.0	0.0	(3,[0],[1.0])	(2,[0],[1.0])	(8,[0,2,3,6],[32....
Indonesia	31	1	Bing	16	1	0.0	2.0	(3,[0],[1.0])	(2,[],[])	(8,[0,1,2,3],[31....
Malaysia	27	1	Google	21	1	3.0	1.0	(3,[],[])	(2,[1],[1.0])	(8,[0,2,7],[27....
Indonesia	29	1	Yahoo	9	0	0.0	0.0	(3,[0],[1.0])	(2,[0],[1.0])	[29.0,1.0,9.0,1.0...
Indonesia	33	1	Yahoo	20	1	0.0	0.0	(3,[0],[1.0])	(2,[0],[1.0])	[33.0,1.0,20.0,1....
Indonesia	35	0	Bing	2	0	0.0	2.0	(3,[0],[1.0])	(2,[],[])	(8,[0,2,3],[35.0,...
India	27	1	Yahoo	21	1	1.0	0.0	(3,[1],[1.0])	(2,[0],[1.0])	[27.0,1.0,21.0,0....

As we can see, now we have one extra column named features, which is nothing but a combination of all the input features represented as a single dense vector.

```
[In]: df.select(['features','Status']).show(10,False)
```

[Out]:

```
+-----------------------------------+------+
|features                           |Status|
+-----------------------------------+------+
|[41.0,1.0,21.0,0.0,1.0,0.0,1.0,0.0]|1     |
|[28.0,1.0,5.0,0.0,0.0,1.0,1.0,0.0] |0     |
|(8,[0,2,5,7],[40.0,3.0,1.0,1.0])   |0     |
|(8,[0,1,2,3],[31.0,1.0,15.0,1.0])  |1     |
|(8,[0,2,7],[32.0,15.0,1.0])        |1     |
|(8,[0,2,5,7],[32.0,3.0,1.0,1.0])   |0     |
|(8,[0,2,5,7],[32.0,6.0,1.0,1.0])   |0     |
|(8,[0,2,3,7],[27.0,9.0,1.0,1.0])   |0     |
|(8,[0,2,3,6],[32.0,2.0,1.0,1.0])   |0     |
|(8,[0,1,2,3],[31.0,1.0,16.0,1.0])  |1     |
+-----------------------------------+------+
```

Let us select only the features column as input and the Status column as output for training the logistic regression model:

```
[In]: model_df=df.select(['features','Status'])
```

We must split the Dataframe into train and test sets in order to train and evaluate the performance of the logistic regression model. We split it in 75/25 ratio and train our model on 75% of the dataset. We can print the shape of train and test data to validate the size:

```
[In]: train_df,test_df=model_df.randomSplit([0.75,0.25])
[In]: print(train_df.count())
[Out]: (14972)

[In]: print(test_df.count())
[Out]: (5028)
```

Let us also validate if the target class is balanced in the train and test sets as well. Otherwise, we have to use some mechanism to maintain the class balance sometimes to improve prediction accuracy:

```
[In]: train_df.groupBy('Status').count().show()
[Out]:
+------+-----+
|Status|count|
+------+-----+
|     1| 7502|
|     0| 7470|
+------+-----+

[In]: test_df.groupBy('Status').count().show()
[Out]:
+------+-----+
|Status|count|
+------+-----+
|     1| 2498|
|     0| 2530|
+------+-----+
```

As we can observe, the class balance is well maintained in both train and test sets. We can now go ahead and build the logistic regression model using features as input and Status as output on the train data:

```
[In]: from pyspark.ml.classification import LogisticRegression

[In]: log_reg=LogisticRegression(labelCol='Status').fit(train_df)

[In]: train_results=log_reg.evaluate(train_df).predictions

[In]: train_results.printSchema()
[Out]:

        root
          |-- features: vector (nullable = true)
          |-- Status: integer (nullable = true)
          |-- rawPrediction: vector (nullable = true)
          |-- probability: vector (nullable = true)
          |-- prediction: double (nullable = false)
```

We can access the predictions made by the model using the evaluate function in Spark, which executes all the steps in an optimized way. That results in another Dataframe that contains four columns in total including prediction and probability. The prediction column signifies the class label the model has predicted for the given row, and the probability column contains two associated probabilities (probability for the negative class at the 0th index and probability for the positive class at the 1st index):

```
[In]: train_results.filter(train_results['Status']==1).filter(train_
results['prediction']==1).select(['Status','prediction','probability']).
show(10,False)
[Out]:
```

```
+------+----------+----------------------------------------------------+
|Status|prediction|probability                                         |
+------+----------+----------------------------------------------------+
|1     |1.0       |[0.23372757081611134,0.7662724291838887]            |
|1     |1.0       |[0.01522593463915585,0.9847740653608441]            |
|1     |1.0       |[1.8849965119321794E-5,0.9999811500348806]          |
|1     |1.0       |[0.13417183733022736,0.8658281626697726]            |
|1     |1.0       |[0.06849296273337586,0.9315070372666241]            |
|1     |1.0       |[0.007793852552826135,0.9922061474471738]           |
|1     |1.0       |[0.0037133402092745806,0.9962866597907254]          |
|1     |1.0       |[9.576735475336262E-6,0.9999904232645247]           |
|1     |1.0       |[0.008340376632806483,0.9916596233671935]           |
|1     |1.0       |[0.008340376632806483,0.9916596233671935]           |
+------+----------+----------------------------------------------------+
```

So, in the preceding results, probability at the 0th index is for Status with value 0, and probability at the 1st index is for Status value of 1. As we can see, in some cases the model is very confident of the target class and predicts status 1 with almost 99% probability. The next part is to check the performance of the model on unseen or test data. We again make use of the evaluate function to make predictions on test data:

```
[In]:results=log_reg.evaluate(test_df).predictions
[
```

```
[In]: results.select(['Status','prediction']).show(10,False)
[Out]:
```

```
+------+----------+
|Status|prediction|
+------+----------+
|0     |0.0       |
|0     |0.0       |
|0     |0.0       |
|0     |0.0       |
|1     |0.0       |
|0     |0.0       |
|1     |1.0       |
|0     |1.0       |
|1     |1.0       |
|1     |1.0       |
+------+----------+
```

As we can observe, there are some cases where the model is misclassifying the target class, whereas at the majority of instances the model is doing a fine job of accurate prediction for both classes.

Confusion Matrix

We will manually create the variables for true positives, true negatives, false positives, and false negatives to calculate the performance metrics of the preceding model on test data:

```
[In]: true_postives = results[(results.Status == 1) & (results.prediction
== 1)].count()
[In]: true_negatives = results[(results.Status == 0) & (results.prediction
== 0)].count()
[In]: false_positives = results[(results.Status == 0) & (results.prediction
== 1)].count()
[In]: false_negatives = results[(results.Status == 1) & (results.prediction
== 0)].count()
```

Accuracy

As discussed already in the chapter, accuracy is the most basic metric for evaluating any classifier. However, it is not the right indicator of the performance of the model due to dependency on the target class balance; but in this case since the class is balanced, this can give a strong sense of overall model performance:

$$\frac{(TP+TN)}{(TP+TN+FP+FN)}$$

```
[In]: accuracy=float((true_postives+true_negatives) /(results.count()))
```

```
[In]:print(accuracy)
```

```
[Out]:0.94272
```

The accuracy of the model that we have built is around 94%.

Recall

Recall rate shows how many of the positive class cases are we able to predict correctly out of the total positive class observations:

$$\frac{TP}{(TP+FN)}$$

```
[In]: recall = float(true_postives)/(true_postives + false_negatives)
```

```
[In]:print(recall)
```

```
[Out]: 0.940
```

The recall rate of the model is around 0.94.

Precision

$$\frac{TP}{(TP+FP)}$$

Precision rate talks about the number of true positives predicted correctly out of all the predicted positive observations:

```
[In]: precision = float(true_postives) / (true_postives + false_positives)
[In]: print(precision)
[Out]: 0.94413
```

So the recall rate and precision rate are also in the same range, which is due to the fact that our target class was well balanced. One thing to notice here is that this could be further improved by tuning the model and finding the most optimal hyperparameters. We will look at model tuning in an upcoming chapter. Now we move on to the last part where we try and put everything together in a pipeline to automate end-to-end prediction. We start by importing Pipeline from PySpark followed by rereading the dataset and splitting it into train and test sets before any kind of feature engineering:

```
[In]: from pyspark.ml import Pipeline

[In]: file_location = "/FileStore/tables/data.csv"
[In]: file_type = "csv"
[In]: infer_schema = "true"
[In]: first_row_is_header = "true"
[In]: delimiter = ","
[In]: new_df = spark.read.format(file_type) \
        .option("inferSchema", infer_schema) \
        .option("header", first_row_is_header) \
        .option("sep", delimiter) \
        .load(file_location)
[In]: train_df,test_df=new_df.randomSplit([0.75,0.25])
```

Now we declare different stages of data handling, feature engineering, and model building into the pipeline. All the stages need to be in the logical order. We start with using StringIndexer followed by OneHotEncoder:

```
[In]: stage_1 = StringIndexer(inputCol= 'Country', outputCol=
'Country_index')
[In]: stage_2 = StringIndexer(inputCol= 'Platform',
outputCol='Platform_index')
```

```
[In]: stage_3 =OneHotEncoder(inputCols=[stage_1.getOutputCol(),stage_2.
getOutputCol()], outputCols= ['Country_vec', 'Platform_vec'])
```

We then declare the next stage, which includes VectorAssembler for creating the feature vector for the model training purpose. In the last stage, we build a logistic regression model on the features built in the previous stage:

```
[In]: stage_4 = VectorAssembler(inputCols=['Age', 'Repeat_Visitor',
'Web_pages_viewed','Country_vec','Platform_vec'],
                              outputCol='features')
[In]: stage_5 = LogisticRegression(featuresCol='features',
labelCol='Status')
```

Now that we have the pipeline ready, we can fit it on the training data and predict the results on the test set:

```
[In]: log_reg_pipeline = Pipeline(stages= [stage_1, stage_2, stage_3,
stage_4, stage_5])
```

```
[In]: model = log_reg_pipeline.fit(train_df)
[In]: train_df = model.transform(train_df)
[In]: test_df=model.transform(test_df)
```

```
[In]: test_df.select('features', 'Status', 'rawPrediction', 'probability',
'prediction').show()
[Out]:
```

```
+--------------------+------+--------------------+--------------------+----------+
|            features|Status|       rawPrediction|         probability|prediction|
+--------------------+------+--------------------+--------------------+----------+
|(8,[0,2,5],[17.0,...|     0|[8.66161455005227...|[0.99982692536488...|       0.0|
|(8,[0,2,5],[17.0,...|     0|[7.91452373819541...|[0.99963473559498...|       0.0|
|(8,[0,2,5],[17.0,...|     0|[7.16743292633855...|[0.99922929426908...|       0.0|
|(8,[0,2,5],[17.0,...|     0|[6.4203421144817,...|[0.99837454725501...|       0.0|
|(8,[0,2,5],[17.0,...|     0|[6.4203421144817,...|[0.99837454725501...|       0.0|
|(8,[0,2,5],[17.0,...|     1|[0.44361561962683...|[0.60912022249821...|       0.0|
|(8,[0,2,5,7],[17....|     0|[9.14306885962621...|[0.99989305285446...|       0.0|
|(8,[0,2,5,7],[17....|     0|[9.14306885962621...|[0.99989305285446...|       0.0|
|(8,[0,2,5,7],[17....|     0|[6.90179642405563...|[0.99899503432777...|       0.0|
|(8,[0,2,5,7],[17....|     0|[6.90179642405563...|[0.99899503432777...|       0.0|
|(8,[0,2,5,7],[17....|     0|[0.92506992920076...|[0.71607401147770...|       0.0|
|(8,[0,2,5,6],[17....|     0|[8.4264940570124,...|[0.99978106003999...|       0.0|
|(8,[0,2,5,6],[17....|     0|[8.4264940570124,...|[0.99978106003999...|       0.0|
|(8,[0,2,5,6],[17....|     0|[6.93231243329868...|[0.99902520922475...|       0.0|
|(8,[0,2,5,6],[17....|     0|[6.18522162144182...|[0.99794458837780...|       0.0|
|(8,[0,2,5,6],[17....|     0|[4.69103999772810...|[0.99090631698484...|       0.0|
|(8,[0,1,2,5],[17....|     1|[-3.5146833390954...|[0.02889732036491...|       1.0|
+--------------------+------+--------------------+--------------------+----------+
```

Conclusion

In this chapter, we went over the process of understanding the building blocks of logistic regression, converting categorical columns into numerical features in PySpark, training the logistic regression model, and automating it using a pipeline.

CHAPTER 6

Random Forests Using PySpark

This chapter will focus on building random forests (RFs) with PySpark for classification. It would also include hyperparameter tuning to find the best set of parameters for the model. We will learn about various aspects of ensembling and how predictions take place, but before knowing more about random forests, we must cover the building block of random forests, which is a decision tree. A decision tree can also be used for classification/regression, but in terms of accuracy, random forests do a better job at predictions due to various reasons, which we will cover later in the chapter. Let's start to learn more about decision trees.

Decision Tree

A decision tree falls under the supervised category of machine learning and uses frequency tables for making predictions. One advantage of a decision tree is that it can handle both categorical and numerical variables. As the name suggests, it operates in sort of a tree structure and forms these rules based on various splits to finally make predictions. The algorithm that is used in a decision tree is ID3 developed by J. R. Quinlan.

We can break down the decision tree in different components as shown in Figure 6-1.

© Pramod Singh 2022
P. Singh, *Machine Learning with PySpark*, https://doi.org/10.1007/978-1-4842-7777-5_6

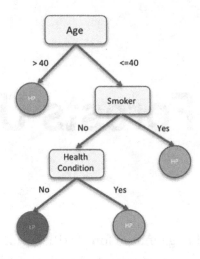

Figure 6-1. *Decision tree*

The topmost split node from where the tree branches out is known as the root node. In the preceding example, Age is the root node. The values represented in circles are known as leaf nodes or predictions. Let's take a sample dataset to understand how a decision tree actually works.

Table 6-1 contains some sample data from people of different age groups and attributes. The final decision to be taken based on these attributes is whether the insurance premium should be on the higher side or not. This is a typical classification case, and we will classify using a decision tree. We have four input columns (Age Group, Smoker, Medical Condition, Salary Level).

Table 6-1. *Example Dataset*

Age Group	Smoker	Medical Condition	Salary Level	Insurance Premium
Old	Yes	Yes	High	High
Teenager	Yes	Yes	Medium	High
Young	Yes	Yes	Medium	Low
Old	No	Yes	High	High
Young	Yes	Yes	High	Low
Teenager	No	Yes	Low	High
Teenager	No	No	Low	Low

(continued)

Table 6-1. (*continued*)

Age Group	Smoker	Medical Condition	Salary Level	Insurance Premium
Old	No	No	Low	High
Teenager	No	Yes	Medium	High
Young	No	Yes	Low	High
Young	Yes	No	High	Low
Teenager	Yes	No	Medium	Low
Young	No	No	Medium	High
Old	Yes	No	Medium	High

Table 6-2. *Entropy Calculation*

Entropy Calculation Target – Insurance Premium Feature – Smoker		Insurance Premium (Target)	
		High (9)	Low (5)
Smoker (Feature)	Yes (7)	3	4
	No (7)	6	1

Entropy

The decision tree makes subsets of this data in such a way that each of those subsets contains same class values (homogenous), and to calculate homogeneity, we use something known as entropy. This can also be calculated using a couple of other metrics like the Gini index and classification error, but we will take up entropy to understand how decision trees work. The formula to calculate entropy is

$$-p \, log_2 p - q \, log_2 q$$

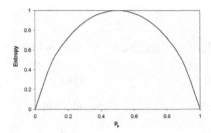

Figure 6-2. *Entropy*

Figure 6-2 shows that entropy is equal to 0 if the subset is completely pure – that means it belongs to only a single class – and it is equal to 1 if the subset is divided equally in two classes.

If we want to calculate entropy of our target variable (Insurance Premium), we have to first calculate probability of each class and then use the preceding formula to calculate entropy.

Insurance Premium	
High (9)	Low (5)

The probability of the High category is equal to 9/14 = 0.64.

The probability of the Low category is equal to 5/14 = 0.36.

$$\text{Entropy} = -p(High)log_2(p(High)) - p(Low)log_2(p(Low))$$

$$= -(0.64 * log_2(0.64)) - (0.36 * log_2(0.36))$$

$$= 0.94$$

In order to build the decision tree, we need to calculate two kinds of entropy:

1. Entropy of target (Insurance Premium)

2. Entropy of target with attribute (e.g., Insurance Premium – Smoker)

We have already seen the entropy of target, so let's calculate the entropy of target with input feature. Let's consider the Smoker feature, for example:

$$Entropy_{(Target, Feature)} = Probability_{Feature} * Entropy_{Categories}$$

$$Entropy_{(Target, Smoker)} = P_{yes} * Entropy_{(3,4)} + P_{no} * Entropy_{(6,1)}$$

$$P_{yes} = \frac{7}{14} = 0.5$$

$$P_{no} = \frac{7}{14} = 0.5$$

$$Entropy_{(3,4)} = -\frac{3}{7} * \log_2\left(\frac{3}{7}\right) - \left(\frac{4}{7}\right) * \log_2\left(\frac{4}{7}\right)$$

$$= 0.99$$

$$Entropy_{(6,1)} = -\frac{6}{7} * \log_2\left(\frac{6}{7}\right) - \left(\frac{1}{7}\right) * \log_2\left(\frac{1}{7}\right)$$

$$= 0.59$$

$$Entropy_{(Target, Smoker)} = 0.55 * 0.99 + 0.5 * 0.59$$

$$= 0.79$$

Similarly, we calculate the entropy for all the other attributes:

$$Entropy_{(Target, Age\ Group)} = 0.69$$

$$Entropy_{(Target, Medical\ Condition)} = 0.89$$

$$Entropy_{(Target, Salary\ Level)} = 0.91$$

Information Gain

Information gain (IG) is used to make the splits in a decision tree. The attribute that offers the maximum information gain is used for splitting the subset. Information gain tells which is the most important feature out of all in terms of making predictions. In terms of entropy, IG is the change in entropy of target before splitting and after splitting on a feature:

$$Information\ Gain = Entropy_{(Target)} - Entropy_{(Target,Feature)}$$

$$IG_{Smoker} = Entropy_{(Target)} - Entropy_{(Target,Smoker)}$$

$$= 0.94 - 0.79$$
$$= 0.15$$

$$IG_{Age\ Group} = Entropy_{(Target)} - Entropy_{(Target,Age\ Group)}$$

$$= 0.94 - 0.69$$
$$= 0.25$$

$$IG_{Medical\ Condition} = Entropy_{(Target)} - Entropy_{(Target,Medical\ Condition)}$$

$$= 0.94 - 0.89$$
$$= 0.05$$

$$IG_{Salary\ Level} = Entropy_{(Target)} - Entropy_{(Target,Salary\ Level)}$$

$$= 0.94 - 0.91$$
$$= 0.03$$

As we can observe, the Age Group attribute gives the maximum information gain. Hence, the decision tree's root node would be Age Group, and the first split happens on that attribute as shown in Figure 6-3.

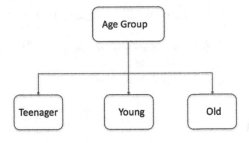

Figure 6-3. *Decision tree split*

The process of finding the next attribute that offers the largest information gain continues recursively, and further splits are made in the decision tree. Finally, the decision tree might look something as shown in Figure 6-4.

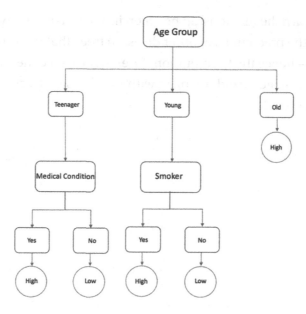

Figure 6-4. *Decision tree splits*

The advantage the decision tree offers is that it can be easily transformed into a set of rules by following the root node to any leaf node and hence can be easily used for classification. There are sets of hyperparameters associated with decision trees that give more options to build trees in a different manner. One of those is max depth, which allows us to decide the depth of the decision tree. The deeper the tree, the more number of splits the tree has and the more chances of overfitting.

Random Forests

Now that we know how a decision tree works, we can move on to random forests. As the name suggests, random forests are made up of many trees, a lot of decision trees. They are quite popular and sometimes are the go-to method for supervised machine learning. Random forests can also be used for classification and regression. They combine votes from a lot of individual decision trees and then predict the class with the majority of votes or take the average in the case of regression. This works really well because the weak learners eventually group together to make strong predictions. The importance lies in the way these decision trees are formed. The name "random" is there for a reason in RF because the trees are formed with a random set of features and a random set of training examples. Now each decision tree being trained of somewhat a different set of

111

data points tries to learn the relationship between input and output, which eventually gets combined with the predictions of other decision trees that use other sets of data points to get trained – hence the term random forests. If we take the similar example as the one earlier and create a random forest with five decision trees, it might look something like in Figure 6-5.

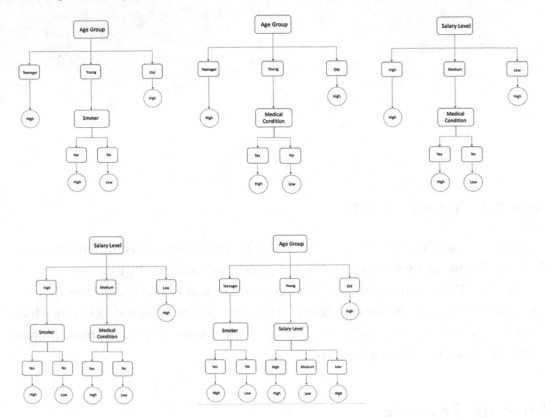

Figure 6-5. *Individual decision trees*

Now, each of these decision trees has used a subset of data to get trained and a subset of features. This is also known as the "bagging" (bootstrap aggregating) technique. Each tree sort of votes regarding the prediction, and the class with the maximum votes is the ultimate prediction by the random forest classifier as shown in Figure 6-6.

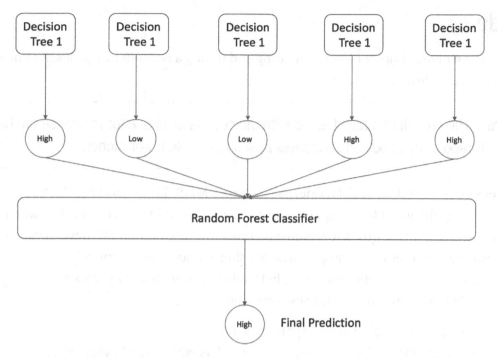

Figure 6-6. *Random forest*

Some of the advantages that random forests offer are mentioned in the following:

- Feature Importance: A random forest can give the importance of each feature that has been used for training in terms of prediction power. This offers great opportunity to select relevant features and drop the weaker ones. The total sum of all feature importance is always equal to 1.

- Increased Accuracy: Since it collects the votes from individual decision trees, the prediction power of random forests is relatively higher compared to a single decision tree.

- Less Overfitting: The results of individual classifiers get averaged or max voted, and hence this reduces the chances of overfitting.

One of the disadvantages of a random forest is that it is difficult to visualize compared to a decision tree and it is a little more on the computation side as it builds multiple individual classifiers.

Code

This section of the chapter focuses on fitting and tuning a random forest classifier using PySpark in Databricks.

Note   The complete dataset along with the code is available for reference on the GitHub repo of this book and executes best on Spark 3.1.0 or higher.

Let's build a random forest model using Spark's MLlib library and predict the target variable using the input features. The dataset that we are going to use for this example is an open source dataset with a few thousand rows and six columns. We have to use five input variables to predict the target variable using a random forest model.

We start the collaborative notebook in Databricks and import pyspark and SparkSession to create a new SparkSession object:

```
[In]: from pyspark.sql import SparkSession
[In]: spark=SparkSession.builder.appName("EnsemblingwithPySpark").
getOrCreate()
```

The next step is to upload the survey .csv data file, which would be used for this part. We follow the same procedure to upload the file in Databricks as we have seen in previous chapters. Feel free to remove old files in case you run into memory issues. In case you are running Spark on a local system, please ensure to have the data present in the same directory where Spark is running or else mention the path where the data resides on your local system. In this case, we read the data from the FileStore location:

```
[In]: file_location = "/FileStore/tables/survey_data.csv"
[In]: file_type = "csv"
[In]: infer_schema = "true"
[In]: first_row_is_header = "true"
[In]: delimiter = ","
[In]: df = spark.read.format(file_type) \
        .option("inferSchema", infer_schema) \
        .option("header", first_row_is_header) \
        .option("sep", delimiter) \
        .load(file_location)

[In]: display(df)
```

[Out]:

	rate_marriage ▲	age ▲	yrs_married ▲	children ▲	religious ▲	affairs ▲	
1	5	32	6	1	3	0	
2	4	22	2.5	0	2	0	
3	3	32	9	3	3	1	
4	3	27	13	3	1	1	
5	4	22	2.5	0	1	1	
6	4	37	16.5	4	3	1	

In this part, we drill deeper into the dataset by viewing the dataset, validating the shape of the dataset, and discussing various statistical measures of the variables. We start with checking the shape of the dataset:

```
[In]:print((df.count(), len(df.columns)))
[Out]: (6366, 6)
```

So the preceding output confirms the size of our dataset, and we can then validate the datatypes of the input values to check if we need to change/cast any column datatypes:

```
[In]: df.printSchema()
```

```
[Out]: root
 |-- rate_marriage: integer (nullable = true)
 |-- age: double (nullable = true)
 |-- yrs_married: double (nullable = true)
 |-- children: double (nullable = true)
 |-- religious: integer (nullable = true)
 |-- affairs: integer (nullable = true)
```

As we can see, there are no categorical columns that need to be converted into numerical form. Let's have a look at the dataset using the show function in Spark:

```
[In]: df.show(5)
```

[Out]:

```
+---------------+----+------------+--------+---------+-------+
|rate_marriage| age|yrs_married|children|religious|affairs|
+---------------+----+------------+--------+---------+-------+
|            5|32.0|        6.0|     1.0|        3|      0|
|            4|22.0|        2.5|     0.0|        2|      0|
|            3|32.0|        9.0|     3.0|        3|      1|
|            3|27.0|       13.0|     3.0|        1|      1|
|            4|22.0|        2.5|     0.0|        1|      1|
+---------------+----+------------+--------+---------+-------+
only showing top 5 rows
```

We can now use the describe function to go over statistical measures of the dataset:

```
[In]: df.describe().select('summary','rate_marriage','age','yrs_married',
'children','religious').show()
[Out]:
```

```
+-------+------------------+------------------+-----------------+------------------+------------------+
|summary|     rate_marriage|               age|      yrs_married|          children|         religious|
+-------+------------------+------------------+-----------------+------------------+------------------+
|  count|              6366|              6366|             6366|              6366|              6366|
|   mean| 4.109644989004084|29.082862079798932| 9.00942507068803|1.3968740182218033|2.4261702796104303|
| stddev|0.9614295945655025| 6.847881883668817|7.280119972766412|1.4334708285603440|0.8783688402641785|
|    min|                 1|              17.5|              0.5|               0.0|                 1|
|    max|                 5|              42.0|             23.0|               5.5|                 4|
+-------+------------------+------------------+-----------------+------------------+------------------+
```

As we can observe, the average age of people is close to 29 years, and the average number of years being married is 9 years.

Let us explore individual columns to understand data in deeper detail. The groupBy function used along with count returns us the frequency of each of the categories in the data:

```
[In]: df.groupBy('affairs').count().show()
[Out]:
```

```
+-------+-----+
|affairs|count|
+-------+-----+
|      1| 2053|
|      0| 4313|
+-------+-----+
```

So we have more than 33% of the people who are involved in some sort of extramarital affairs out of the total number of people.

```
[In]: df.groupBy('rate_marriage').count().show()
[Out]:
```

116

```
+--------------+-----+
|rate_marriage|count|
+--------------+-----+
|            1|   99|
|            3|  993|
|            5| 2684|
|            4| 2242|
|            2|  348|
+--------------+-----+
```

The majority of the people rate their marriage very high (4 or 5), and the rest rate it on the lower side. Let's drill down a little bit further to understand if marriage rating is related to the affairs variable or not:

```
[In]: df.groupBy('rate_marriage','affairs').count().orderBy('rate_marriage',
'affairs','count',ascending=True).show()
[Out]:
```

```
+--------------+-------+-----+
|rate_marriage|affairs|count|
+--------------+-------+-----+
|            1|      0|   25|
|            1|      1|   74|
|            2|      0|  127|
|            2|      1|  221|
|            3|      0|  446|
|            3|      1|  547|
|            4|      0| 1518|
|            4|      1|  724|
|            5|      0| 2197|
|            5|      1|  487|
+--------------+-------+-----+
```

Clearly, the figures indicate the high percentage of people having affairs rated their marriages low. This might prove to be a useful feature for the prediction. We will explore other variables as well in similar manner:

```
[In]: df.groupBy('religious','affairs').count().orderBy('religious',
'affairs','count',ascending=True).show()
[Out]:
```

```
+----------+-------+-----+
|religious|affairs|count|
+----------+-------+-----+
|         1|      0|  613|
|         1|      1|  408|
|         2|      0| 1448|
|         2|      1|  819|
|         3|      0| 1715|
|         3|      1|  707|
|         4|      0|  537|
|         4|      1|  119|
+----------+-------+-----+
```

We have a similar story from the religious perspective. A higher percentage of affair involvement can be seen on the people who have rated lower on the religious feature.

117

```
[In]: df.groupBy('children','affairs').count().orderBy('children','affairs',
'count',ascending=True).show()
[Out]:
```

```
+--------+-------+-----+
|children|affairs|count|
+--------+-------+-----+
|     0.0|      0| 1912|
|     0.0|      1|  502|
|     1.0|      0|  747|
|     1.0|      1|  412|
|     2.0|      0|  873|
|     2.0|      1|  608|
|     3.0|      0|  460|
|     3.0|      1|  321|
|     4.0|      0|  197|
|     4.0|      1|  131|
|     5.5|      0|  124|
|     5.5|      1|   79|
+--------+-------+-----+
```

The preceding table does not clearly indicate any of the trends regarding relation between number of children and chances of being involved in affairs. Let us use the groupBy function along with mean to know more about the dataset:

```
[In]: df.groupBy('affairs').mean().show()
[Out]:
```

```
+-------+------------------+-----------------+-----------------+------------------+------------------+------------+
|affairs|avg(rate_marriage)|         avg(age)| avg(yrs_married)|     avg(children)| avg(religious)|avg(affairs)|
+-------+------------------+-----------------+-----------------+------------------+------------------+------------+
|      1|3.6473453482708234|30.537018996590355|11.152459814905017|1.7289332683877252| 2.261568436434486|         1.0|
|      0| 4.329700904242986| 28.39067934152562| 7.989334569904939|1.2388128912589844|2.5045212149316023|         0.0|
+-------+------------------+-----------------+-----------------+------------------+------------------+------------+
```

So the people who have affairs rate their marriage low and are a little on the higher side from the age standpoint. They are also married for a higher number of years and less religious. This is the part where we create a single vector combining all input features by using Spark's VectorAssembler:

```
[In]: from pyspark.ml.feature import VectorAssembler
```

We need to assemble all of the input columns into a single vector that would act as the input feature for the model. So we select the input columns that we need to use to create the single feature vector and name the output vector as features:

```
[In]: df_assembler = VectorAssembler(inputCols=['rate_marriage', 'age',
'yrs_married', 'children', 'religious'], outputCol="features")

[In}:df = df_assembler.transform(df)
```

118

```
[In]: df.printSchema()
[Out]:
root
 |-- rate_marriage: integer (nullable = true)
 |-- age: double (nullable = true)
 |-- yrs_married: double (nullable = true)
 |-- children: double (nullable = true)
 |-- religious: integer (nullable = true)
 |-- affairs: integer (nullable = true)
 |-- features: vector (nullable = true)
```

As we can see, now we have one extra column named features, which is nothing but a combination of all the input features represented as a single dense vector:

```
[In]: df.select(['features','affairs']).show(10,False)
[Out]:
```

```
+---------------------------+-------+
|features                   |affairs|
+---------------------------+-------+
|[5.0,32.0,6.0,1.0,3.0]     |0      |
|[4.0,22.0,2.5,0.0,2.0]     |0      |
|[3.0,32.0,9.0,3.0,3.0]     |1      |
|[3.0,27.0,13.0,3.0,1.0]    |1      |
|[4.0,22.0,2.5,0.0,1.0]     |1      |
|[4.0,37.0,16.5,4.0,3.0]    |1      |
|[5.0,27.0,9.0,1.0,1.0]     |1      |
|[4.0,27.0,9.0,0.0,2.0]     |1      |
|[5.0,37.0,23.0,5.5,2.0]    |1      |
|[5.0,37.0,23.0,5.5,2.0]    |1      |
+---------------------------+-------+
only showing top 10 rows
```

Let us select only the features column as input and the affairs column as output for training the random forest model:

```
[In]: model_df=df.select(['features','affairs'])
```

We have to split the dataset into training and test sets in order to train and evaluate the performance of the random forest model. We split it in 75/25 ratio and train our model on 75% of the dataset. We can print the shape of train and test data to validate the size:

```
[In]: train_df,test_df=model_df.randomSplit([0.75,0.25])
[In]: print(train_df.count())
[Out]:4775
```

```
[In]: train_df.groupBy('affairs').count().show()
[Out]:
```

```
+-------+-----+
|affairs|count|
+-------+-----+
|      1| 1530|
|      0| 3288|
+-------+-----+
```

This ensures we have a balanced set of values for the target class ("affairs") into training and test sets:

```
[In]: test_df.groupBy('affairs').count().show()
```

```
[Out]:
```

```
+-------+-----+
|affairs|count|
+-------+-----+
|      1|  523|
|      0| 1025|
+-------+-----+
```

In this part, we build and train the random forest model using features as input and affairs as output. We will also try to find the best set of hyperparameters through cross-validation. We use ParamGridBuilder in PySpark to declare a set of hyperparameters and make a threefold cross-validation for model training and hyperparameter tuning. You can choose to alter the numFolds parameter in cross-validation as per your choice. There are many hyperparameters that can be set to tweak the performance of the model, but we are choosing a couple of them here (number of trees and maximum depth):

```
[In]: from pyspark.ml.classification import RandomForestClassifier
```

```
[In]: from pyspark.ml.tuning import ParamGridBuilder, CrossValidator
```

```
[In]: model=RandomForestClassifier(labelCol='affairs')
```

```
[In]: paramGrid = ParamGridBuilder()\
        .addGrid(model.numTrees,[5,10,20,25,30,40,50])\
        .addGrid(model.maxDepth,[2,3,5,6,8,9,10])\
        .build()
```

```
[In]: from pyspark.ml.evaluation import BinaryClassificationEvaluator
[In]:evaluator=BinaryClassificationEvaluator(rawPredictionCol=
'rawPrediction',labelCol='affairs')
```

```
[In]: cv = CrossValidator(estimator=model, estimatorParamMaps=paramGrid,
evaluator=evaluator, numFolds=3)
```

```
[In]: rf_model = cv.fit(train_df)
```

Now that we have fit the model on train data, we can test its performance on test data. The first column in the predictions table is that of input features of the test data. The second column is the actual label or output of the test data. The third column (rawPrediction) represents the measure of confidence for both possible outputs. The fourth column is of conditional probability for each class label, and the final column is the prediction by the random forest classifier. We can apply a groupBy function on the prediction column to find out the number of predictions made for the positive and the negative class:

```
[In]: test_results=rf_model.transform(test_df)
[In]: test_results.show()
```

features	affairs	rawPrediction	probability	prediction
[1.0,22.0,2.5,0.0...	1	[14.6041967294583...	[0.29208393458916...	1.0
[1.0,22.0,2.5,0.0...	1	[16.0932303205154...	[0.32186460641030...	1.0
[1.0,22.0,2.5,1.0...	0	[17.7239032353726...	[0.35447806470745...	1.0
[1.0,22.0,2.5,1.0...	0	[19.2192402721879...	[0.38438480544375...	1.0
[1.0,27.0,2.5,0.0...	0	[14.2152260900801...	[0.28430452180160...	1.0
[1.0,27.0,6.0,0.0...	0	[18.8525524550372...	[0.37705104910074...	1.0
[1.0,27.0,6.0,1.0...	1	[18.3786805465211...	[0.36757361093042...	1.0
[1.0,27.0,6.0,2.0...	1	[19.3152479691891...	[0.38630495938378...	1.0
[1.0,27.0,9.0,4.0...	0	[20.9219018279125...	[0.41843803655825...	1.0
[1.0,32.0,13.0,2....	1	[15.2094265653290...	[0.30418853130658...	1.0
[1.0,32.0,13.0,2....	1	[12.9702263358626...	[0.25940452671725...	1.0
[1.0,32.0,16.5,3....	1	[17.1442313409021...	[0.34288462681804...	1.0
[1.0,37.0,13.0,3....	1	[16.0227955310337...	[0.32045591062067...	1.0
[1.0,37.0,16.5,1....	1	[15.2566244058027...	[0.30513248811605...	1.0
[1.0,37.0,16.5,2....	1	[15.8784129457800...	[0.31756825891560...	1.0
[1.0,37.0,16.5,3....	1	[12.6530379071666...	[0.25306075814333...	1.0
[1.0,37.0,16.5,3....	1	[12.6530379071666...	[0.25306075814333...	1.0
[1.0,42.0,16.5,2....	1	[16.1127125117274...	[0.32225425023454...	1.0
[1.0,42.0,16.5,5....	1	[22.7022609214829...	[0.45404521842965...	1.0
[1.0,42.0,23.0,2....	1	[15.9138711184069...	[0.31827742236813...	1.0

```
[In]: test_results.groupBy('prediction').count().show()
[Out]:
```

```
+----------+-----+
|prediction|count|
+----------+-----+
|       0.0| 1267|
|       1.0|  281|
+----------+-----+
```

Next, we make use of BinaryClassEvaluator and MulticlassClassificationEvaluator to look at the performance of the model on test data:

```
[In]: from pyspark.ml.evaluation import BinaryClassificationEvaluator
[In]:rf_auc=BinaryClassificationEvaluator(labelCol='affairs').
evaluate(test_results)
[In]: print(rf_auc)
[Out]: 0.7397584293242548
```

```
[In]: from pyspark.ml.evaluation import MulticlassClassificationEvaluator
[In]:rf_accuracy=MulticlassClassificationEvaluator(labelCol='affairs',
metricName='accuracy').evaluate(test_results)
[In]: print(rf_accuracy)
[Out]: 0.7093023255813954
```

As we can observe, the model prediction accuracy is around 71% in this case. In order to select the model with best parameters, we can use bestModel as shown in the following:

```
[In]: bestModel = rf_model.bestModel
```

As mentioned in the earlier part, RF gives the importance of each feature in terms of predictive power, and it is very useful to figure out the critical variables that contribute the most in model predictions:

```
[In]: bestModel.featureImportances
[Out]: SparseVector(5, {0: 0.5341, 1: 0.0481, 2: 0.2601, 3: 0.0731, 4:
0.0847})
```

We used five features, and the importance can be found out using the feature importance function. To know which input feature is mapped to which index value, we can use metadata information:

```
[In]: df.schema["features"].metadata["ml_attr"]["attrs"]
```

```
[Out]:
[{'idx': 0, 'name': 'rate_marriage'},
{'idx': 1, 'name': 'age'},
{'idx': 2, 'name': 'yrs_married'},
{'idx': 3, 'name': 'children'},
{'idx': 4, 'name': 'religious'}]}
```

So rate_marriage is the most important feature from a prediction standpoint followed by yrs_married. The least significant variable seems to be age. The next step is to save this best-performing model and reload it for predictions again on test data. We will create a new folder in Databricks to persist the trained model object:

```
[In]: from pyspark.ml.classification import RandomForestClassificationModel
```

```
[In]: bestModel.save("/FileStore/tables/model/final_model")
In]:rf=RandomForestClassificationModel.load("/FileStore/tables/model/
final_model")
[In]: model_preditions=rf.transform(test_df)
[In]: model_preditions.show()
[Out]:
```

features	affairs	rawPrediction	probability	prediction
[1.0,22.0,2.5,1.0...]	1	[20.6275737300354...]	[0.51568934325088...]	0.0
[1.0,22.0,2.5,1.0...]	0	[21.0517688362509...]	[0.52629422090627...]	0.0
[1.0,22.0,2.5,1.0...]	1	[21.0517688362509...]	[0.52629422090627...]	0.0
[1.0,27.0,2.5,0.0...]	1	[16.6484348341049...]	[0.41621087085262...]	1.0
[1.0,27.0,2.5,0.0...]	1	[16.6484348341049...]	[0.41621087085262...]	1.0
[1.0,27.0,6.0,2.0...]	1	[12.5149665043640...]	[0.31287416260910...]	1.0
[1.0,27.0,6.0,3.0...]	0	[10.3688151987548...]	[0.25922037996887...]	1.0
[1.0,27.0,9.0,1.0...]	1	[12.7561919594105...]	[0.31890479898526...]	1.0
[1.0,27.0,9.0,4.0...]	0	[11.5143613102822...]	[0.28785903275705...]	1.0
[1.0,32.0,2.5,1.0...]	0	[18.7834222723131...]	[0.46958555680782...]	1.0
[1.0,32.0,13.0,0....]	1	[15.5233424414447...]	[0.38808356103611...]	1.0
[1.0,32.0,13.0,2....]	1	[11.0675324990557...]	[0.27668831247639...]	1.0
[1.0,32.0,13.0,3....]	1	[13.3355951902811...]	[0.33338987975702...]	1.0
[1.0,32.0,16.5,5....]	0	[9.2352685732684,...]	[0.23088171433170...]	1.0
[1.0,37.0,13.0,3....]	1	[14.2413311172001...]	[0.35603327793000...]	1.0
[1.0,37.0,13.0,5....]	0	[14.9544158155306...]	[0.37386039538826...]	1.0
[1.0,37.0,16.5,3....]	1	[10.1053087509594...]	[0.25263271877398...]	1.0

As we can see, we loaded the pre-trained model in PySpark and used it to predict on test data. The next and last part of this chapter is to combine all the steps seen previously and run end to end using Spark pipelines. We reload the survey data and create a new dataframe:

```
[In]: file_location = "/FileStore/tables/survey_data.csv"
[In]: file_type = "csv"
[In]: infer_schema = "true"
[In]: first_row_is_header = "true"
[In]: delimiter = ","
[In]: new_df = spark.read.format(file_type) \
        .option("inferSchema", infer_schema) \
        .option("header", first_row_is_header) \
        .option("sep", delimiter) \
        .load(file_location)

[In]: display(new_df)

[Out]:
```

	rate_marriage	age	yrs_married	children	religious	affairs
1	5	32	6	1	3	0
2	4	22	2.5	0	2	0
3	3	32	9	3	3	1
4	3	27	13	3	1	1
5	4	22	2.5	0	1	1
6	4	37	16.5	4	3	1

The next step is to split the data intro training and test sets followed by declaring the different stages of the pipeline. Since there were no categorical columns in the dataframe this time, we can directly call VectorAssembler in the first stage:

```
[In]: train_df,test_df=new_df.randomSplit([0.75,0.25])
```

```
[In]: from pyspark.ml import Pipeline
```

```
[In]:stage_1 = VectorAssembler(inputCols=['rate_marriage', 'age','yrs_
married', 'children', 'religious'],outputCol='features')
```

The next step is to build a random forest model on the training dataframe and make the predictions on the test set:

```
[In]:stage_2 =RandomForestClassifier(featuresCol='features',labelCol='affairs')
```

```
[In]: pipeline = Pipeline(stages= [stage_1, stage_2])
```

```
[In]: model = pipeline.fit(train_df)
```

```
[In]: train_df = model.transform(train_df)
```

```
[In]: test_df=model.transform(test_df)
```

```
[In]:test_df.select('features', 'affairs', 'rawPrediction', 'probability',
'prediction').show()
[Out]:
```

```
+--------------------+-------+--------------------+--------------------+----------+
|            features|affairs|       rawPrediction|         probability|prediction|
+--------------------+-------+--------------------+--------------------+----------+
|[1.0,17.5,0.5,0.0...|      0|[16.0165087908187...|[0.80082543954093...|       0.0|
|[1.0,22.0,2.5,1.0...|      0|[7.96634181262584...|[0.39831709063129...|       1.0|
|[1.0,27.0,2.5,0.0...|      1|[8.16592333915508...|[0.40829616695775...|       1.0|
|[1.0,27.0,6.0,1.0...|      0|[7.03784331351195...|[0.35189216567559...|       1.0|
|[1.0,27.0,9.0,2.0...|      1|[5.60797549032096...|[0.28039877451604...|       1.0|
|[1.0,27.0,9.0,4.0...|      0|[6.24860049032096...|[0.31243002451604...|       1.0|
|[1.0,27.0,13.0,2....|      1|[5.47109228980579...|[0.27355461449028...|       1.0|
|[1.0,32.0,13.0,1....|      1|[5.65875209337976...|[0.28293760466898...|       1.0|
|[1.0,32.0,13.0,3....|      1|[6.03095258720335...|[0.30154762936016...|       1.0|
|[1.0,32.0,13.0,3....|      1|[6.03095258720335...|[0.30154762936016...|       1.0|
|[1.0,32.0,16.5,2....|      1|[6.03095258720335...|[0.30154762936016...|       1.0|
|[1.0,32.0,16.5,3....|      1|[5.30798921963035...|[0.26539946098151...|       1.0|
|[1.0,32.0,16.5,3....|      1|[6.03095258720335...|[0.30154762936016...|       1.0|
|[1.0,32.0,16.5,3....|      1|[6.03095258720335...|[0.30154762936016...|       1.0|
|[1.0,32.0,16.5,5....|      0|[6.64286650698326...|[0.33214332534916...|       1.0|
|[1.0,37.0,13.0,3....|      0|[10.0591546603395...|[0.50295773301697...|       0.0|
|[1.0,37.0,13.0,5....|      0|[7.41662624972314...|[0.37083131248615...|       1.0|
```

Conclusion

In this chapter, we went over the process of understanding the building blocks of random forests and creating a ML model in PySpark for classification along with hyperparameter tuning. We also covered the steps to save the trained ML model and reuse it for predictions.

CHAPTER 7

Clustering in PySpark

So far, we have seen supervised Machine Learning where the target variable or label is known to us, and we try to predict the output based on the input features. Unsupervised indicates that there is no labeled data and we don't try to predict any output. Instead, we try to find interesting patterns and come up with groups within the data. It's more of an art rather than going after the prediction accuracy. The values within the groups are very similar to each other, whereas any two groups are very distinct from each other. Let's take an example to understand clustering.

When we join a new school or college, we come across many new faces, and everyone looks so different. We hardly know anyone in the institute, and there are no groups in place initially. Slowly and gradually, we start spending time with other people, and the groups start to develop. We interact with a lot of different people and figure out how similar and dissimilar they are to us. A few months down the lane, we are almost settled in our own groups of friends. These can be small groups or moderate-size groups sometimes, but each group is different from another group in some ways. The friends/ members within the group have similar attributes/likings/tastes and hence stay together. Clustering is somewhat similar to this approach of forming groups based on a set of attributes, which define the groups.

We can apply clustering on any sort of that where we want to form groups of similar observations and use it for better decision-making. In early days, customer segmentation used to be done by the rule-based approach, which was much of a manual effort and could only use a limited number of variables. For example, if businesses want to do customer segmentation, they will consider up to ten variables such as age, gender, salary, location, etc. and create rule-based segments, which still gave reasonable performance. However, in today's scenario, that would become highly ineffective. One reason is data availability in abundance, and the other is dynamic customer behavior. There are thousands of other variables that can be considered to come up with these machine learning–driven segments, which are more enriched and meaningful.

© Pramod Singh 2022
P. Singh, *Machine Learning with PySpark*, https://doi.org/10.1007/978-1-4842-7777-5_7

When we start clustering, each observation is different and doesn't belong to any group; but based on how similar the attributes of each observation are, we group them in such a way that each group contains most similar records and there is as much difference as possible between any two groups. So how do we measure if two observations are similar or different?

There are multiple approaches to calculate the distance between any two observations. Primarily, we represent any observation as a form of vector that contains the value of that particular observation (A) as shown in the following:

Age	Salary ($'0000)	Weight (Kgs)	Height(Ft.)
32	8	65	6

Now, let's say we want to calculate the distance of this observation/record from any other observation (B), which also contains similar attributes as shown in the following:

Age	Salary ($'000)	Weight (Kgs)	Height(Ft.)
40	15	90	5

We can measure the distance using the Euclidean method, which is pretty straightforward. It is defined as the distance between two points as the square root of the sum of the squares of the differences between the corresponding coordinates of the points. It is also known as the Cartesian distance. We are trying to calculate the distance of a straight line between any two points, and if the distance between those points is small, they are more likely to be similar, whereas if the distance is large, they are dissimilar to each other as shown in Figure 7-1.

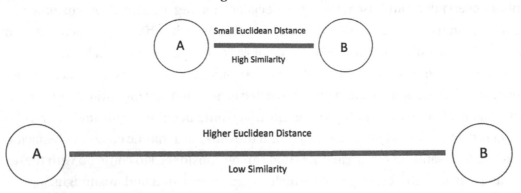

Figure 7-1. *Euclidean distance–based similarity*

The Euclidean distance between any two points can be calculated using the following formula:

dist(A,B) = $\sqrt{(A1-B1)^2 + (A2-B2)^2 + (A3-B3)^2 + (A4-B4)^2}$

dist(A,B)= $\sqrt{((Age\ diff)^2 + (Salary\ diff)^2 + (Weight\ diff)^2 + (Height\ diff)^2))}$

dist(A,B)= $\sqrt{((32-40)^2 + (8-15)^2 + (65-90)^2 + (6-5)^2)}$

dist(A,B)= $\sqrt{(64+49+625+1)}$

dist(A,B) = 27.18

Hence, the Euclidean distance between observations A and B is 27.18. The other techniques to calculate the distance between observations are

1. Manhattan distance

2. Mahalanobis distance

3. Minkowski distance

4. Chebyshev distance

5. Cosine distance

The aim of clustering is to have minimal intra-cluster distance and maximum inter-cluster difference. We can end up with different groups based on the distance approach that we have used to do clustering, and hence it's critical to be sure of opting for the right distance metric that aligns with the business problem. Before going into different clustering techniques, let's quickly go over some of the applications of clustering.

Applications

Clustering is used in a variety of use cases these days ranging from customer segmentation to anomaly detection. Businesses widely use machine learning–driven clustering for profiling customers and segmentation to create market strategies around these results. Clustering drives a lot of search engine results by finding similar objects in one cluster and dissimilar objects far from each other. It recommends the nearest similar result based on the search query.

Clustering can be done in multiple ways based on the type of data and business requirement. The most used ones are K-means and hierarchical clustering.

K-Means

"K" stands for the number of clusters or groups that we want in a given dataset. This type of clustering involves deciding on the number of clusters in advance. Before looking at how K-means clustering works, let's get familiar with a couple of terms first:

1. Centroid

2. Variance

Centroid refers to the center data point at the center of a cluster or a group. It is also the most representative point within the cluster as it's the utmost equidistant data point from the other points within the cluster. The centroid (represented by a cross) for three random clusters is shown in Figure 7-2.

Figure 7-2. *Centroids of clusters*

Based on the changes in data points in clusters, the centroid values also change. The center position within the group alters resulting in a new centroid as shown in Figure 7-3.

Figure 7-3. *New centroids of new clusters*

The whole idea of clustering is to minimize intra-cluster distance, that is, internal distance of data points from the centroid of the cluster, and maximize inter-cluster distance, that is, between centroids of two different clusters.

Variability is the total sum of intra-cluster distances between the centroid and data points within that cluster as shown in Figure 7-4. Variability keeps on decreasing with increase in the number of clusters. The more the number of clusters, the lesser the number of data points within each cluster and hence the lesser the variability.

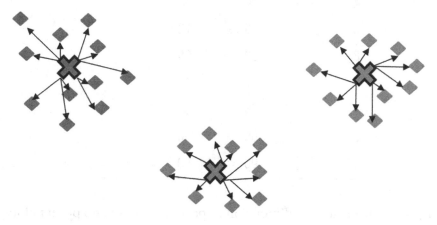

Figure 7-4. *Intra-cluster distance*

K-means clustering is composed of four steps in total to form the internal groups within the dataset. We will consider a sample dataset to understand how the K-means clustering algorithm works. The dataset contains a few users with their age and weight values as shown in Table 7-1. Now we will use K-means clustering to come up meaningful clusters and understand the algorithm.

Table 7-1. *Sample Dataset for K-Means*

User ID	Age	Weight
1	18	80
2	40	60
3	35	100
4	20	45
5	45	120
6	32	65
7	17	50
8	55	55
9	60	90
10	90	50

If we plot these users in two-dimensional space, we can see no point belongs to any group initially, and our intention is to find clusters (may be two or three) within this group of users such that each group contains similar users. Each user is represented by the age and weight as shown in Figure 7-5.

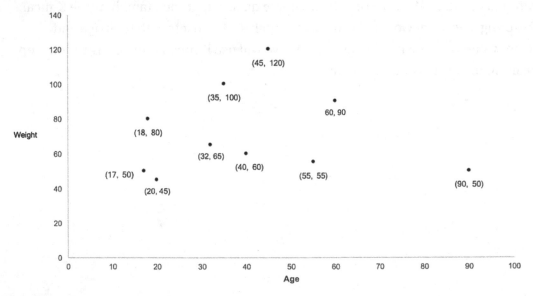

Figure 7-5. *Users before clustering*

Step 1: Deciding on K

It starts with deciding on the number of clusters (K). Most of the times, we are not sure of the right number of groups at the start, but we can find the best number of clusters using a method called as the elbow method based on variability. For this example, let's start with K = 2 to keep things simple. So we are looking for two clusters within this sample data.

Step 2: Random Initialization of Centroids

The next step is to randomly consider any two of the points to be the centroids of the new clusters. These can be chosen randomly. Hence, we select user number 5 and user number 10 as the two centroids on the new clusters as shown in Table 7-2.

Table 7-2. *Sample Dataset for K-Means*

User ID	Age	Weight
1	18	80
2	40	60
3	35	100
4	20	45
5 (Centroid 1)	45	120
6	32	65
7	17	50
8	55	55
9	60	90
10 (centroid 2)	90	50

The centroids can be represented by weight and age values as shown in Figure 7-6.

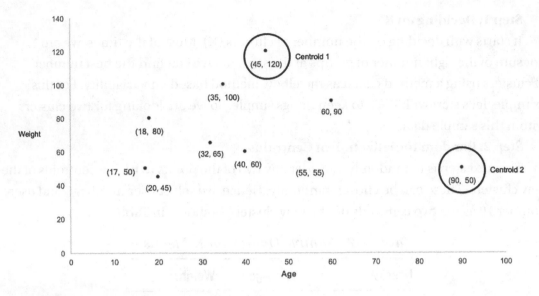

Figure 7-6. *Random centroids of two clusters*

Step 3: Assigning Cluster Number to Each Value

In this step, we calculate the distance of each point from the centroids. In this example, we calculate the Euclidean squared distance of each user from the two centroid points. Based on the distance value, we go ahead and decide which particular cluster the user belongs to (1 or 2). Whichever centroid the user is near to (less distance), the user would become part of that cluster. The Euclidean squared distance is calculated for each user as shown in Table 7-3. The distance of user 5 and user 10 would be zero from respective centroids as they are the same points as the centroids.

Table 7-3. *Cluster Assignment Based on Distance from Centroids*

User ID	Age	Weight	ED* from Centroid 1	ED* from Centroid 2	Cluster
1	18	80	48	78	1
2	40	60	60	51	2
3	35	100	22	74	1
4	20	45	79	70	2
5	45	120	0	83	1
6	32	65	57	60	1
7	17	50	75	73	2
8	55	55	66	35	2
9	60	90	34	50	1
10	90	50	83	0	2

(*Euclidean distance)

So, as per the distance from centroids, we have allocated each user to either cluster 1 or cluster 2. Cluster 1 contains five users, and cluster 2 also contains five users. The initial clusters are shown in Figure 7-7.

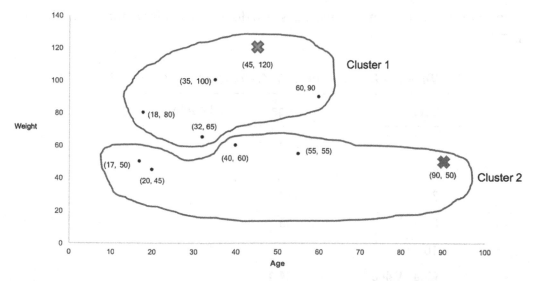

Figure 7-7. *Initial clusters and centroids*

As discussed earlier, the centroids of the clusters are bound to change after inclusion or exclusion of new data points in the clusters. As the earlier centroids (C1, C2) are no longer at the center of the clusters, we calculate new centroids in the next step.

Step 4: Calculating New Centroids and Reassigning Clusters

The final step in K-means clustering is to calculate the new centroids of the clusters and reassign the clusters to each value based on the distance from the new centroids. Let's calculate the new centroids of cluster 1 and cluster 2. To calculate the centroid of cluster 1, we simply take the mean of age and weight for only those values that belong to cluster 1 as shown in Table 7-4.

Table 7-4. *New Centroid Calculation of Cluster 1*

User ID	Age	Weight
1	18	80
3	35	100
5	45	120
6	32	65
9	60	90
Mean Value	**38**	**91**

The centroid calculation for cluster 2 is also done in a similar manner and shown in Table 7-5.

Table 7-5. *New Centroid Calculation of Cluster 2*

User ID	Age	Weight
2	40	60
4	20	45
7	17	50
8	55	55
10	90	50
Mean Value	**44.4**	**52**

Now we have new centroid values for each cluster represented by a cross as shown in Figure 7-8. The arrow signifies the movement of the centroid within the cluster.

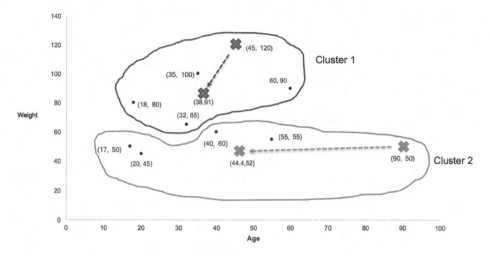

Figure 7-8. *New centroids of both clusters*

With the centroids of each cluster, we repeat step 3 of calculating the Euclidean squared distance of each user from the new centroids and finding out the nearest centroid. We then reassign the users to either cluster 1 or cluster 2 based on the distance from the centroids. In this case, only one value (user 6) changes its cluster from 1 to 2 as shown in Table 7-6.

Table 7-6. *Reallocation of Clusters*

User ID	Age	Weight	ED* from Centroid 1	ED* from Centroid 2	Cluster
1	18	80	23	38	1
2	40	60	31	9	2
3	35	100	9	49	1
4	20	45	49	25	2
5	45	120	30	68	1
6	32	65	27	18	2
7	17	50	46	27	2
8	55	55	40	11	2
9	60	90	22	41	1
10	90	50	66	46	2

Now, cluster 1 is left with only four users, and cluster 2 contains six users based on the distance from each cluster's centroid as shown in Figure 7-9.

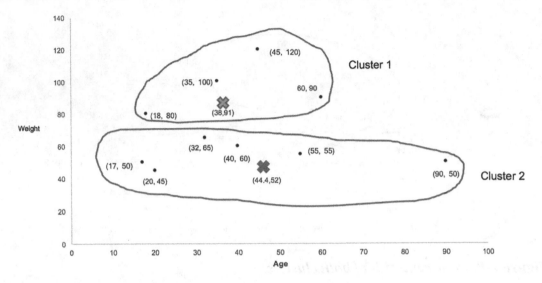

Figure 7-9. *Reallocation of clusters*

We keep repeating the preceding steps till there is no more change in cluster allocations. The centroids of new clusters are shown in Table 7-7.

Table 7-7. *Calculation of Centroids*

User ID	Age	Weight
1	18	80
3	35	100
5	45	120
9	60	90
Mean Value	39.5	97.5
User ID	**Age**	**Weight**
2	40	60
4	20	45
6	32	65

(*continued*)

Table 7-7. (*continued*)

User ID	Age	Weight
7	17	50
8	55	55
10	90	50
Mean Value	42.33	54.17

As we go through the steps, the centroid movements keep becoming small, and values almost become part of that particular cluster as shown in Figure 7-10.

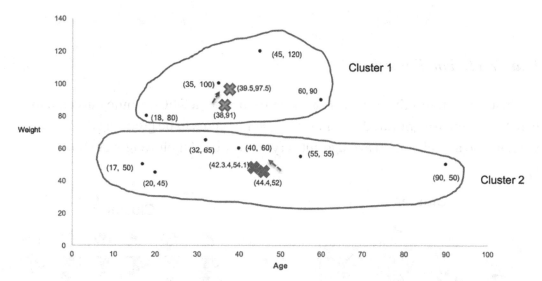

Figure 7-10. *Reallocation of clusters*

As we can observe, there is no more change in the points even after change in centroids. That completes K-means clustering. The results can vary as it's based on the first set of random centroids. To reproduce the results, we can set the starting points ourselves as well. The final clusters with values are shown in Figure 7-11.

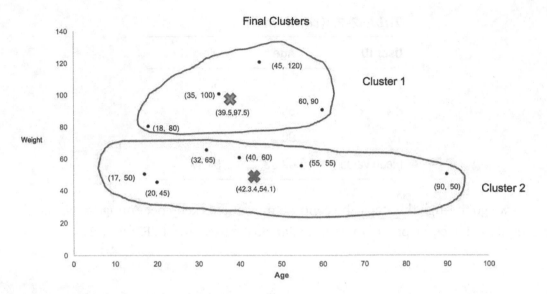

Figure 7-11. *Final clusters*

Cluster 1 contains the users who are average on the height attribute but seem to be higher on the weight variable, whereas cluster 2 seems to be grouping those users together who are taller than average but very conscious of their weight as shown in Figure 7-12.

Figure 7-12. *Attributes of final clusters*

Deciding on the Number of Clusters (K)

Selecting an optimal number of clusters is quite tricky most of the time as we need a deep understanding of the dataset and the context of the business problem. Additionally, there is no right or wrong answer when it comes to unsupervised learning. One approach might result in a different number of clusters compared with another approach. We have to try and figure out which approach works the best and if the clusters created are relevant enough for decision-making. Each cluster can be represented with a few important attributes, which signify or give information about that particular cluster. However, there is a method to pick the best possible number of clusters with a dataset. This method is known as the elbow method.

Elbow Method

This method helps us to measure the total variability in the data with the number of clusters. The more number of clusters, the lesser the variability. If we have an equal number of clusters to the number of records in the dataset, then the variability would be zero because the distance of each point from itself is zero. The variability or SSE (sum of squared errors) along with "K" values is shown in Figure 7-13.

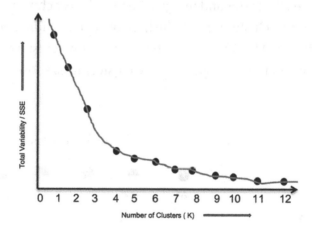

Figure 7-13. *Elbow method*

As we can observe, there is some sort of elbow formation between K values of 3 and 4. There is sudden reduction in total variability (intra-cluster difference), and the variability sort of declines very slowly after that. In fact, it flattens after the K = 9 value. So the value of K = 3 makes the most sense if we go with the elbow method as it captures the most variability with the less number of clusters.

Hierarchical Clustering

This is another type of unsupervised machine learning technique and is different from K-means in the sense that we don't have to know the number of clusters in advance. There are two types of hierarchical clustering:

1. Agglomerative clustering (bottom-up approach)

2. Divisive clustering (top-down approach)

Agglomerative Clustering

This type of clustering starts with the assumption that each data point is a separate cluster and gradually keeps combining the nearest values into same clusters till all the values become part of one cluster. This is a bottom-up approach that calculates the distance between each cluster and merges the two closest clusters into one. Let's understand agglomerative clustering with the help of visualization. Let's say we have seven data points initially (A1–A7) that need be grouped into clusters where similar values are grouped together using agglomerative clustering as shown in Figure 7-14.

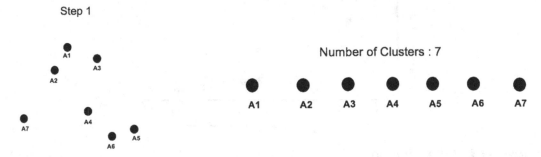

Figure 7-14. *Each value as individual cluster*

At the initial stage (step 1), each point is treated as an individual cluster. In the next step, the distance between every point is calculated, and the nearest points are combined together into a single cluster. In this example, A1 and A2 and A5 and A6 are nearest to each other and hence form a single cluster as shown in Figure 7-15.

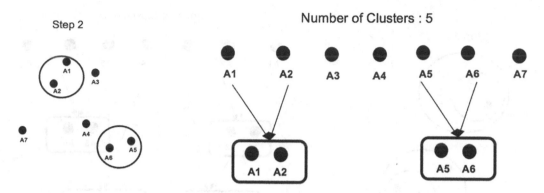

Figure 7-15. *Nearest clusters merged together*

Deciding on the most optimal number of clusters while using hierarchical clustering can be done in multiple ways. One way is to use the elbow method itself, and the other option is by making use of something known as a dendrogram. It is used to visualize the variability between clusters (Euclidean distance). In a dendrogram, the height of the vertical lines represents the distance between points or clusters and data points listed along the bottom. Each point is plotted on the x-axis, and the distance is represented on the y-axis (length). It is the hierarchical representation of the data points. In this example, the dendrogram at step 2 looks like as shown in Figure 7-16.

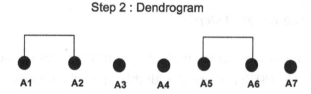

Figure 7-16. *Dendrogram*

In step 3, the exercise of calculating distance between clusters is repeated, and the nearest clusters are combined into a single cluster. This time A3 gets merged with (A1, A2) and A4 with (A5, A6) as shown in Figure 7-17.

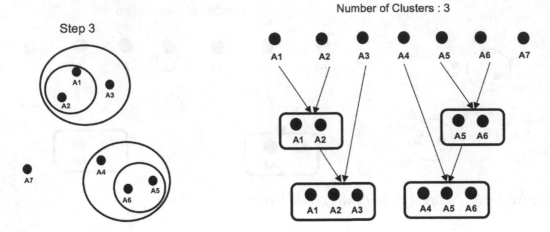

Figure 7-17. *Nearest clusters merged together*

The dendrogram after step 3 is shown in Figure 7-18.

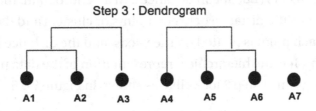

Figure 7-18. *Dendrogram post step 3*

In step 4, the distance of the only remaining point A7 gets calculated, and it is found nearer to cluster (A4, A5, A6). It is merged with this same cluster as shown in Figure 7-19.

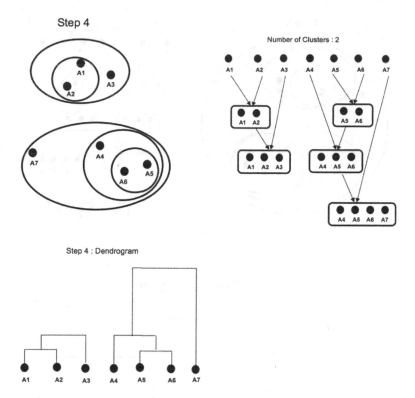

Figure 7-19. *Cluster formation*

At the last stage (step 5), all the points get combined into a single cluster (A1, A2, A3, A4, A5, A6, A7) as shown in Figure 7-20.

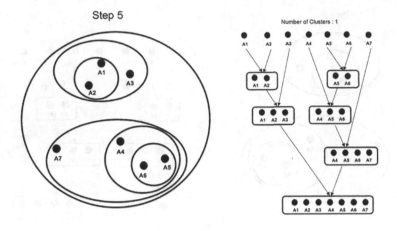

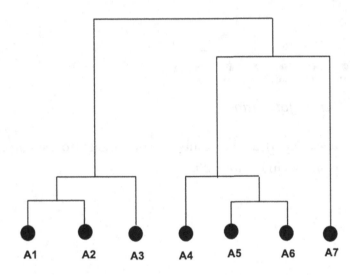

Final Dendrogram

Figure 7-20.   *Agglomerative clustering*

Sometimes it is difficult to identify the right number of clusters by the dendrogram as it can become very complicated and difficult to interpret depending on the dataset being used to do clustering. Hierarchical clustering doesn't work well on large datasets compared with K-means. Clustering is also very sensitive to the scale of data points. Hence, it's always advised to do data scaling before clustering. There are other types of clustering that can be used to group similar data points together such as

1. Gaussian mixture model clustering

2. K-modes clustering

3. Fuzzy C-means clustering

But the preceding methods are beyond the scope of this book. We now jump into using a dataset for building clusters using K-means in PySpark.

Code

This section of the chapter covers K-means clustering using PySpark visualizing the clusters in a Databricks notebook.

Note   The complete dataset along with the code is available for reference on the GitHub repo of this book and executes best on Spark 3.1.0.

For this exercise, we consider the most standardized open source dataset out there – Iris dataset – to capture the cluster number and compare supervised and unsupervised performance. You could also choose to try another dataset for the clustering purpose.

Data Info

The dataset contains a total of 150 records with 5 columns (sepal length, sepal width, petal length, petal width, species). There are 50 records for each type of species. We will try to group these into clusters without using the species label information.

We start the Databricks notebook and import pyspark and SparkSession and create a new SparkSession object as we have been doing in previous chapters:

```
[In]: import pyspark
[In]: from pyspark.sql import SparkSession
[In]: spark=SparkSession.builder.appName('K_means').getOrCreate()
```

The next step is to upload the Iris .csv data file in Databricks as we have seen in previous chapters. Again, feel free to remove old files in case you run into memory issues:

```
[In]: file_location = "/FileStore/tables/iris_dataset.csv"
[In]: file_type = "csv"
[In]: infer_schema = "true"
[In]: first_row_is_header = "true"
[In]: delimiter = ","
[In]: df = spark.read.format(file_type) \
      .option("inferSchema", infer_schema) \
      .option("header", first_row_is_header) \
      .option("sep", delimiter) \
      .load(file_location)

[In]: display(df)
[Out]:
```

	sepal_length ▲	sepal_width ▲	petal_length ▲	petal_width ▲	species ▲
1	5.1	3.5	1.4	0.2	setosa
2	4.9	3	1.4	0.2	setosa
3	4.7	3.2	1.3	0.2	setosa
4	4.6	3.1	1.5	0.2	setosa
5	5	3.6	1.4	0.2	setosa
6	5.4	3.9	1.7	0.4	setosa

Figure 7-21. *Iris dataset*

Now we explore the dataset further and validate the shape and its as shown in Figure 7-21. The output confirms the size of our dataset, and we can then validate the datatypes of the input values to check if we need to change/cast any column datatypes:

```
[In]:print((df.count(), len(df.columns)))
[Out]: (150,5)

[In]: df.printSchema()
[Out]: root
 |-- sepal_length: double (nullable = true)
```

```
|-- sepal_width: double (nullable = true)
|-- petal_length: double (nullable = true)
|-- petal_width: double (nullable = true)
|-- species: string (nullable = true)
```

There are a total of five columns, out of which four are numerical; the label column is categorical:

```
[In]: from pyspark.sql.functions import rand, randn

[In]: df.orderBy(rand()).show(10,False)
[Out]:
+------------+-----------+------------+-----------+----------+
|sepal_length|sepal_width|petal_length|petal_width|species   |
+------------+-----------+------------+-----------+----------+
|5.5         |2.6        |4.4         |1.2        |versicolor|
|4.5         |2.3        |1.3         |0.3        |setosa    |
|5.1         |3.7        |1.5         |0.4        |setosa    |
|7.7         |3.0        |6.1         |2.3        |virginica |
|5.5         |2.5        |4.0         |1.3        |versicolor|
|6.3         |2.3        |4.4         |1.3        |versicolor|
|6.2         |2.9        |4.3         |1.3        |versicolor|
|6.3         |2.5        |4.9         |1.5        |versicolor|
|4.7         |3.2        |1.3         |0.2        |setosa    |
|6.1         |2.8        |4.0         |1.3        |versicolor|
+------------+-----------+------------+-----------+----------+

[In]: df.groupBy('species').count().orderBy('count').show(10,False)
[Out]:
+----------+-----+
|species   |count|
+----------+-----+
|virginica |50   |
|setosa    |50   |
|versicolor|50   |
+----------+-----+
```

So it confirms that there are an equal number of records for each species available in the dataset. This is the part where we create a single vector combining all input features by using Spark's VectorAssembler. It creates only a single feature that captures the input values for a particular row. So, instead of four input columns (we are not considering the label column since it's an unsupervised machine learning technique), we essentially translate it into a single column with four input values in the form of a list:

```
[In]: from pyspark.ml.linalg import Vector
[In]: from pyspark.ml.feature import VectorAssembler

[In]: input_cols=['sepal_length', 'sepal_width', 'petal_length',
'petal_width']

[In]: vec_assembler = VectorAssembler(inputCols = input_cols,
outputCol='features')

[In]:final_data = vec_assembler.transform(df)
```

Step 5: Building the K-Means Clustering Model

The final data contains the input vector that can be used to run K-means clustering. Since we need to declare the value of "K" in advance before using K-means, we can use the elbow method or silhouette coefficient method to figure out the right value of "K." In order to use the elbow method, we run K-means clustering for different values of "K." First, we import KMeans from the PySpark library and create an empty list that would capture the variability or SSE (within cluster distance) for each value of K:

Note   computeCost to calculate the sum of squared errors is deprecated in Spark 3.0.1, and ClusteringEvaluator is used to calculate cluster distance.

```
[In]: from pyspark.ml.clustering import KMeans
[In]: errors=[]

[In]: for k in range(2,10):
        kmeans = KMeans(featuresCol='features',k=k)
        model = kmeans.fit(final_data)
        intra_distance = model.summary.trainingCost
        errors.append(intra_distance)
```

```
print("With K={}".format(k))
print("Within Set Sum of Squared Errors = " + str(intra_distance))
print('--'*30)
```

Note   We make use of trainingCost in order to access the sum of squared distances in clusters.

Now, we can plot the intra-cluster distance with the number of clusters using NumPy and matplotlib:

```
[In]: import pandas as pd
[In]: import numpy as np
[In]: import matplotlib.pyplot as plt
[In]: cluster_number = range(2,10)
[In]: plt.xlabel('Number of Clusters (K)')
[In]: plt.ylabel('SSE')
[In]: plt.scatter(cluster_number,errors)
[In]: plt.show()
```

[Out]:

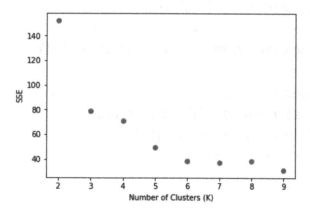

In this case, k = 3 seems to be the best number of clusters as we can see a sort of elbow formation between the 3 and 4 values. We will build the final clusters using k = 3, but before that we can also explore the ClusteringEvaluator method, which provides the

151

silhouette coefficient. You could choose this method as well to decide on the number of clusters. The key thing to remember regarding the silhouette coefficient is that it ranges between –1 and 1 similar to a correlation coefficient:

1. The value of 1 indicates that the item is far away from the neighboring clusters.

2. The value of 0 indicates that the item is on or very close to the decision boundary between two neighboring clusters.

3. The value less than 0 indicates that there is overlap and clusters are not proper.

```
[In]: from pyspark.ml.evaluation import ClusteringEvaluator
[In]: silhouette_score=[]

[In]: for k in range(2,10):
        kmeans = KMeans(featuresCol='features',k=k)
        model = kmeans.fit(final_data)
        # Make predictions
        predictions = model.transform(final_data)
        evaluator = ClusteringEvaluator()
        silhouette = evaluator.evaluate(predictions)
        silhouette_score.append(silhouette)
        print("With K={}".format(k))
        print("silhouette Score  = " + str(silhouette))
        print('--'*30)
[In]: cluster_number = range(2,10)
[In]: plt.plot(cluster_number,silhouette_score)
[In]: plt.xlabel('Number of Clusters (K)')
[In]: plt.ylabel('silhouette_score')
[In]: plt.show()
[Out]:
```

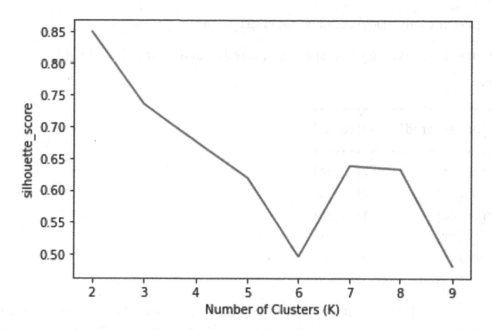

Now that we have seen the silhouette-based clustering evaluation as well, we can build the final clusters using K-means with value of K being 3:

```
[In]: kmeans = KMeans(featuresCol='features',k=3)
[In]: model = kmeans.fit(final_data)
[In]: model.transform(final_data).groupBy('prediction').count().show()
[Out]:
```

```
+----------+-----+
|prediction|count|
+----------+-----+
|         1|   50|
|         2|   38|
|         0|   62|
+----------+-----+
```

K-means clustering gives us three different clusters based on the Iris dataset. We certainly made few of the allocations wrong as only one category has got 50 records in the group and the rest of the categories are mixed up. We can use the transform function to assign the cluster number to the original dataset and use a groupBy function to validate the groupings:

```
[In]: predictions=model.transform(final_data)

[In]: predictions.groupBy('species','prediction').count().show()

[Out]:
+----------+----------+-----+
|   species|prediction|count|
+----------+----------+-----+
| virginica|         2|   14|
|    setosa|         0|   50|
| virginica|         1|   36|
|versicolor|         1|    3|
|versicolor|         2|   47|
+----------+----------+-----+
```

As it can be observed, the setosa species is perfectly grouped along with versicolor almost being captured in the same cluster, but virginica seems to fall within two different groups. K-means can produce different results every time as it chooses the starting point (centroid) randomly every time. Hence, the results that you might get in your K-means clustering might be totally different from these results.

In the final step, we can visualize the new clusters with the help of Python's matplotlib library. In order to do that, we convert our Spark Dataframe into a Pandas dataframe first:

```
[In]: pandas_df = predictions.toPandas()
[In]: pandas_df.head()
```

	sepal_length	sepal_width	petal_length	petal_width	species	features	prediction
5	5.4	3.9	1.7	0.4	setosa	[5.4, 3.9, 1.7, 0.4]	0
95	5.7	3.0	4.2	1.2	versicolor	[5.7, 3.0, 4.2, 1.2]	2
132	6.4	2.8	5.6	2.2	virginica	[6.4, 2.8, 5.6, 2.2]	1
128	6.4	2.8	5.6	2.1	virginica	[6.4, 2.8, 5.6, 2.1]	1
23	5.1	3.3	1.7	0.5	setosa	[5.1, 3.3, 1.7, 0.5]	0

We import required libraries to plot the 3D visualization and observe the clusters:

```
[In]: from mpl_toolkits.mplot3d import Axes3D
```

```
[In]: cluster_vis = plt.figure(figsize=(12,10)).gca(projection='3d')
```

```
[In]: cluster_vis.scatter(pandas_df.sepal_length, pandas_df.sepal_width,
pandas_df.petal_length, c=pandas_df.prediction,depthshade=False)
```

```
[In]: plt.show()
```

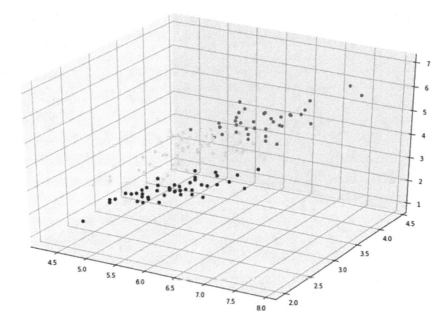

Conclusion

In this chapter, we went over different types of unsupervised machine learning techniques and also built clusters using K-means algorithms in PySpark. We also covered various techniques to decide on the optimal number of clusters to have meaningful clusters based on the sample data.

CHAPTER 8

Recommender Systems

In brick-and-mortar stores, we have salespeople guiding and recommending us the relevant products while shopping. On the other hand, on online retail platforms, there are zillions of different products available, and we have to navigate ourselves to find the right product. The situation is that users have too many options and choices available, yet don't like to invest a lot of time going through the entire catalogue of items. Hence, the role of recommender systems (RSs) becomes critical for recommending relevant items and driving customer conversion.

Traditional physical stores use a planogram to arrange the items in such a way that can increase the visibility of high-selling items and can increase revenue, whereas online retail stores need to keep it dynamic based on the preference of each individual customer rather than keeping it the same for everyone.

Recommender systems are mainly used for auto-suggesting the right content or product to the right user in a personalized manner to enhance the overall experience. Recommender systems are really powerful in terms of using huge amount of data and learning to understand the preference for specific users. Recommendations help users to easily navigate through millions of products or tons of content (articles/videos/movies) and show them the right item/information that they might like or buy. So, in simple terms, a RS helps discover information on behalf of the users. Now, it depends on the users to decide if the RS did a good job at recommendations or not, and they can choose to either select the product/content or discard and move on. Each of the decisions of users (positive or negative) helps to retrain the RS on latest data to be able to give even better recommendations. In this chapter, we will go over how the RS works and what are different types of techniques used under the hood for making these recommendations. We will also build a recommender system using PySpark.

© Pramod Singh 2022

P. Singh, *Machine Learning with PySpark*, https://doi.org/10.1007/978-1-4842-7777-5_8

Recommendations

Recommender systems can be used for multiple purposes in the sense of recommending various things to the users. For example, some of them might fall in the following categories:

1. Retail products

2. Jobs

3. Connections/friends

4. Movies/music/videos/books/articles

5. Ads

The "What to Recommend" part totally depends on the context in which the RS is used and can help the business to increase revenues by providing most likely items that users can buy or increasing the engagement by showcasing relevant content at the right time. The RS takes care of the critical aspect that the product or content that is being recommended should be something that users might like but would not have discovered on their own. Along with that, the RS also needs an element of varied recommendations to keep it interesting enough. Examples of heavy usage of RSs by businesses today are Amazon products, Facebook's friend suggestions, LinkedIn's "People You May Know," Netflix's movies, YouTube's videos, Spotify's music, and Coursera's courses.

The impact of these recommendations is proving to be immense from a business standpoint, and hence more time is being spent in making these RSs more efficient and relevant. Some of the immediate benefits that RSs offer in retail settings are

1. Increased revenue

2. Positive reviews and ratings by users

3. Increased engagement

For the other verticals such as ad recommendations and other content recommendations, the RS helps immensely to help them find the right thing for users and hence increase adoptions and subscriptions. Without the RS, recommending online content to millions of users in a personalized manner or offering generic content to each user can be incredibly off target and lead to negative impact on users.

Now that we know the usage and features of RSs, we can take a look at different types of RSs. There are mainly five types of RSs that can be built:

1. Popularity-based RS

2. Content-based RS

3. Collaborative filtering (CF)–based RS

4. Hybrid RS

5. Association rule mining–based RS

We will briefly go over each one of these except for the last item, that is, association rule mining–based RS, as it's out of scope for this book.

Popularity-Based RS

This is the most basic and simplest RS that can be used to recommend products or content to the users. It recommends items/content based on items/content bought/viewed/liked/downloaded by most of the users. While it is easy and simple to implement, it doesn't produce relevant results as the recommendations stay the same for every user, but it sometimes outperforms some of the more sophisticated RSs. The way this RS is implemented is by simply ranking the items on various parameters and recommending the top-ranked items in the list. As already mentioned, items or content can be ranked by

1. No. of times downloaded

2. No. of times bought

3. No. of times viewed

4. Highest rated

5. No. of times shared

6. No. of times liked

This kind of RS directly recommends the best-selling or most watched/bought items to the customers and hence increases the chances of customer conversion. The limitation of this RS is that it is not hyper-personalized.

Content-Based RS

This type of RS recommends similar items to the user that the user has liked in the past. So the whole idea is to calculate a similarity score between any two items and recommend to the user based upon the profile of the user's interests. We start with creating item profiles for each of the items. Now these item profiles can be created in multiple ways, but the most common approach is to include information regarding the details or attributes of the item. For example, the item profile of a movie can have values on various attributes such as Horror, Art, Comedy, Action, Drama, and Commercial, as shown in the following:

Movie ID	Horror	Art	Comedy	Action	Drama	Commercial
2310	0.01	0.3	0.8	0.0	0.5	0.9

The preceding table is an example of an item profile, and each of the items would have a similar vector representing its attributes. Now, let's assume the user has watched ten such movies and really liked them. So, for that particular user, we end with an item matrix as given in Table 8-1.

Table 8-1. *Movie Attributes*

Movie ID	Horror	Art	Comedy	Action	Drama	Commercial
2310	0.01	0.3	0.8	0.0	0.5	0.9
2631	0.0	0.45	0.8	0.0	0.5	0.65
2444	0.2	0.0	0.8	0.0	0.5	0.7
2974	0.6	0.3	0.0	0.6	0.5	0.3
2151	0.9	0.2	0.0	0.7	0.5	0.9
2876	0.0	0.3	0.8	0.0	0.5	0.9
2345	0.0	0.3	0.8	0.0	0.5	0.9
2309	0.7	0.0	0.0	0.8	0.4	0.5
2366	0.1	0.15	0.8	0.0	0.5	0.6
2388	0.0	0.3	0.85	0.0	0.8	0.9

User Profile

The other component in a content-based RS is the user profile, which is created using item profiles that the user has liked or rated. Assuming that the user has liked the movies in Table 8-1, the user profile might look like a single vector that is simply the mean of item vectors. The user profile might look something like what is shown in Table 8-2.

Table 8-2. *User Profile*

User ID	Horror	Art	Comedy	Action	Drama	Commercial
1A92	0.251	0.23	0.565	0.21	0.52	0.725

This approach to create the user profile is the most baseline one, and there are other sophisticated ways to create more enriched user profiles such as normalized values, weighted values, etc. The next step is to recommend the items (movies) that this user might like based on the earlier preferences. So the similarity score between the user profile and an item profile is calculated and ranked accordingly. The higher the similarity score, the higher the chances of the movie being liked by the user. There are a couple of ways by which the similarity score can be calculated.

Euclidean Distance

The user profile and item profile both are high-dimensional vectors, and hence to calculate the similarity between two, we need to calculate the distance between both vectors. The Euclidean distance can be easily calculated for n-dimensional vectors using the following formula:

$$d(x,y) = \sqrt{(x_1 - y_n)^2 + \cdots + (x_n - y_n)^2}$$

The higher the distance value, the less similar are the two vectors. Hence, the distance between the user profile and all other items is calculated and ranked in decreasing order. The top few items are recommended to the user in this manner.

Cosine Similarity

Another way to calculate the similarity score between the user and item profiles is cosine similarity. Instead of distance, it measures the angle between two vectors (user profile vector and item profile vector). The smaller the angle between both vectors, the more similar they are to each other. Cosine similarity can be found out using the following formula:

$$\text{sim}(x,y) = \cos(\theta) = x*y / |x|*|y|$$

Let's go over some of the pros and cons of a content-based RS.

Advantages:

1. A content-based RS works independent of other users' data and hence can be applied to an individual's historical data.

2. The rationale behind this RS can be easily understood as the recommendations are based on the similarity score between the user profile and item profiles.

3. New and unknown items can also be recommended to users just based on historical interests and preferences of users.

Disadvantages:

1. An item profile can be biased and might not reflect exact attribute values, which might lead to incorrect recommendations.

2. Recommendations entirely depend on the history of the user, and hence recommend items are similar to historically watched/liked items, meaning new interests or likings of the visitor are not taken into consideration.

Collaborative Filtering–Based RS

A CF-based RS doesn't require the item attributes or description for recommendations but instead works on user item interactions. These interactions can be measured in various ways such as ratings, item bought, time spent, shared on another platform, etc. Before diving deep in CF, let's take a step back and reflect on how we make certain decisions on a day-to-day basis, decisions such as

1. Which movie to watch

2. Which book to read

3. Which restaurant to go to

4. Which place to travel to

We ask our friends, right? We ask for recommendations from people who are similar to us in some ways and have same tastes and likings as ours. Our interests match in some areas, and hence we trust their recommendations. These people can be our family members, friends, colleagues, relatives, or community members. In real life, it's easy to know who are the people falling in this circle; but when it comes to online recommendations, the key task in collaborative filtering is to find the users who are most similar to you. Each user can be represented by a vector that contains the feedback value of user item interaction. Let's understand the user item (UI) matrix first to understand the CF approach.

User Item Matrix

A user item matrix is exactly what the name suggests. In the rows, we have all the unique users, and along the columns we have all the unique items. The values are filled with a feedback or interaction score to highlight the liking or disliking of the user for that product. A simple user item matrix might look something as shown in Table 8-3.

Table 8-3. *User Item Matrix*

User ID	Item 1	Item 2	Item 3	Item 4	Item 5	Item n
14SD	1	4			5	
26BB		3	3			1
24DG	1	4	1		5	2
59YU		2			5	
21HT	3	2	1	2	5	
68BC		1				5
26DF	1	4		3	3	
25TR	1	4			5	
33XF	5	5	5	1	5	5
73QS	1		3			1

As you can observe, the user item matrix is generally very sparse as there are millions of items and each user doesn't interact with every item. Hence, the matrix contains a lot of null values. The values in the matrix are generally feedback values deduced based upon interaction of users with particular items. There are two types of feedback that can be considered in a UI matrix.

Explicit Feedback

This sort of feedback is generally when a user gives a rating to an item after the interaction and experiencing the item features. Ratings can be of multiple types:

1. Rating on a 1–5 scale

2. Simple rating through recommending to others (yes or no or never)

3. Liked the item (yes or no)

Explicit feedback data contains a very limited amount of data points as a very small percentage of users take out time to give ratings even after buying or using the item. A perfect example can be of a movie, as very few users give the ratings even after watching the movie. Hence, building RSs solely on explicit feedback data can put us in a tricky situation. Although the data itself is less noisy, sometimes it's not enough to build recommender systems.

Implicit Feedback

This kind of feedback is not direct and mostly inferred from the activities of the user on the online platform and is based on interactions with items. For example, if the user has bought the item, added to cart, viewed, or spent a great deal of time on information of the item, this indicates that the user has a higher amount of interest in the item. Implicit feedback values are easy to collect, and plenty of data points are available for each user as they navigate their way through the online platform. The challenge with implicit feedback is that it contains a lot of noisy data and hence doesn't add too much value in the recommendations.

Now that we understand the UI matrix and the type of values that go into that matrix, we can see the different types of collaborative filtering. There are mainly two kinds of CF:

1. Nearest neighbors–based CF

2. Latent factor–based CF

Nearest Neighbors–Based CF

This CF works by finding out the K-nearest neighbors of a user by finding the most similar users who also like or dislike the same items as the active user (for the user we are trying to recommend to). There are two steps involved in nearest neighbors collaborative filtering. The first step is to find the K-nearest neighbors, and the second step is to predict the rating or likelihood of whether the active user would like a particular item. The K-nearest neighbors can be found out using some of the earlier techniques we have discussed in the chapter. Metrics such as cosine similarity and Euclidean distance can help us to find the most similar users to the active user out of the total number of users based on the common items both groups have liked or disliked. One other metric that can also be used is Jaccard similarity. Let's take an example to understand this metric. Going back to the earlier user item matrix and taking just five users, the data is as shown in Table 8-4.

Table 8-4. *User Item Matrix*

User ID	Item 1	Item 2	Item 3	Item 4	Item 5	Item n
14SD	1	4			5	
26BB		3	3			1
24DG	1	4	1		5	2
59YU		2			5	
26DF	1	4		3	3	

Let's say we have in total five users and we want to find the two nearest neighbors to the active user (14SD). Jaccard similarity can be found out using

$$sim(x, y) = |Rx \cap Ry| / |Rx \cup Ry|$$

So this means the number of items that any two users have rated in common divided by the total number of items both users have rated.

sim (user1, user2) = 1 / 5 = 0.2 (Since they have rated only Item 2 in common)

The similarity score for the rest of the four users with the active user then would look something as shown in Table 8-5.

Table 8-5. *User Similarity Score*

User ID	Similarity Score
14SD	1
26BB	0.2
24DG	0.6
59YU	0.677
26DF	0.75

So, according to Jaccard similarity, the top two nearest neighbors are the fourth and fifth users. There is a major issue with this approach though. Jaccard similarity doesn't consider the feedback value while calculating the similarity score and only considers the common items rated. So there could be a possibility that users might have rated many items in common but one might have rated them high and the other might have rated them low. The Jaccard similarity score still might end up high for both users, which is counterintuitive. In the preceding example, it is clearly evident that the active user is most similar to the third user (24DG) as they have exact same ratings for three common items, but the third user doesn't even appear in the top two nearest neighbors. Hence, we can opt for other metrics to calculate the K-nearest neighbors.

Missing Values

The user item matrix would contain a lot of missing values for the simple reason that there are a lot of items and not every user interacts with each item. There are a couple of ways to deal with missing values in a UI matrix:

1. Replace the missing values with 0s.

2. Replace the missing values with average ratings of the user.

The more similar the ratings on common items, the nearer the neighbor is to the active user. There are again two categories of nearest neighbors–based CF:

1. User-based CF

2. Item-based CF

The only difference between both RSs is that in user-based CF we find K-nearest users and in item-based CF we find K-nearest items to be recommended to users. We will see how recommendations work in a user-based RS.

As the name suggests, in user-based CF, the whole idea is to find the most similar user to the active user and recommend the items that the similar user has bought/rated highly to the active user, which they haven't seen/bought/tried yet. The assumption that this kind of RS makes is that if two or more users have the same opinion about a bunch of items, then they are likely to have the same opinion about other items as well. Let's take an example to understand user-based collaborative filtering. There are three users, one of which we want to recommend a new item to. The two users are the top two nearest neighbors in terms of likes and dislikes of items with the active user as shown in Figure 8-1.

Nearest Neighbors   **Active User**

Figure 8-1. *Active user and nearest neighbors (images used are from iconfinder.com and under the commercial usage license)*

All three users have rated a particular camera brand very highly, and the first two users are the most similar users to the active user based on their similarity scores as shown in Figure 8-2.

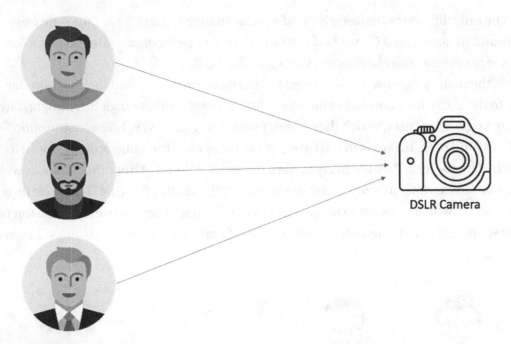

Figure 8-2. *All users like an item (images used are from iconfinder.com and under the commercial usage license)*

Now, the first two users have also rated another item (Xbox 360) very highly, which the third user is yet to interact with and has not seen yet as shown in Figure 8-3. Using this information, we try to predict the rating that the active user would give to the new item (Xbox 360), which again is the weighted average of ratings of the nearest neighbors for that particular item (Xbox 360).

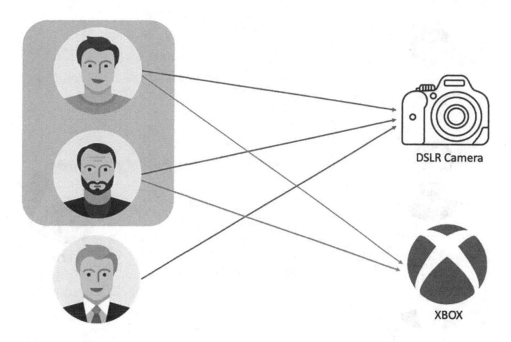

Figure 8-3. *Nearest neighbors also like another item (images used are from iconfinder.com and under the commercial usage license)*

User-based CF then recommends the other item (Xbox 360) to the active user since they are most likely to rate this item higher as the nearest neighbors have rated this item highly as shown in Figure 8-4.

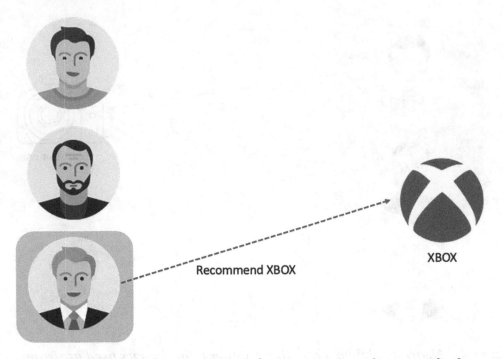

Figure 8-4. *Active user recommendation (images used are from iconfinder.com and under the commercial usage license)*

Latent Factor–Based CF

This kind of collaborative filtering also uses a user item matrix, but instead of finding nearest neighbors and predicting ratings, it tries to decompose the UI matrix into two latent factor matrices. The latent factors are derived values from original values. They are intrinsically related to the observed variables. These new matrices are much lower in terms of rank and contain latent factors. This is also known as matrix factorization. Let's take an example to understand the matrix factorization process. We can decompose an mxn-size matrix "A" of rank r into two lower-rank matrices X and Y such that the dot product of X and Y results in the original A matrix. Let's say we have a matrix A as shown in Table 8-6.

Table 8-6. *Sample Matrix*

1	2	3	5
2	4	8	12
3	6	7	13

We can write all the column values as linear combinations of the first and third columns (A1 and A3):

$$A1 = 1 * A1 + 0 * A3$$

$$A2 = 2 * A1 + 0 * A3$$

$$A3 = 0 * A1 + 1 * A3$$

$$A4 = 2 * A1 + 1 * A3$$

Now we can create the two lower-rank matrices in such a way that the product between those two would return the original matrix A:

X =

1	3
2	8
3	7

Y =

1	2	0	2
0	0	1	1

X contains columns values of A1 and A3, and Y contains the coefficients of linear combinations.

The dot product between X and Y results back in matrix "A" (original matrix).

Considering the same user item matrix as shown in Table 1-2, we factorize or decompose it into two lower-rank matrices:

1. Users latent factor matrix

2. Items latent factor matrix

Table 8-7. User Item Matrix

User ID	Item 1	Item 2	Item 3	Item 4	Item 5	Item n
14SD	1	4			5	
26BB		3	3			1
24DG	1	4	1		5	2
59YU		2			5	
21HT	3	2	1	2	5	
68BC		1				5
26DF	1	4		3	3	
25TR	1	4			5	
33XF	5	5	5	1	5	5
73QS	1		3			1

Table 8-8. Users Latent Factor Matrix

User ID	USF1	USF2	USF3
14SD	0.02	0.97	
26BB		0.24	0.65
24DG	0.03	0.86	0.07
59YU		0.45	
21HT	0.65	0.38	0.05
68BC		0.03	
26DF	0.02	0.78	
25TR	0.01	0.84	
33XF	0.95	0.98	0.93
73QS	0.03		0.48

Table 8-9. Items Latent Factor Matrix

	Item 1	Item 2	Item 3	Item 4	Item 5	Item n
ITF1	0.3	0.23			0.9	
ITF2		0.1	0.14			0.02
ITF3	0.25	0.8	0.09		0.9	0.33

$$\frac{\begin{matrix} 0.23 \\ 0.1 \\ 0.8 \end{matrix}}{}$$

The users latent factor matrix contains all the users mapped to these latent factors, and similarly the items latent factor matrix contains all items in columns mapped to each of the latent factors. The process of finding these latent factors is done using machine learning optimization techniques such as alternating least squares (ALS). The user item matrix is decomposed into latent factor matrices in such a way that the user rating for any item is the product between the user latent factor value and item latent factor value. The main objective is to minimize the total sum of squared errors over the entire user item matrix ratings and predicted item ratings. For example, the predicted rating of the second user (26BB) for Item 2 would be

Rating (user2, item2) =

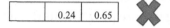

There would be some amount of error on each of the predicted ratings, and hence the cost function becomes the overall sum of squared errors between predicted ratings and actual ratings. Training the recommendation model includes learning these latent factors in such a way that it minimizes the SSE for overall ratings. We can use the ALS method to find the lowest SSE. The way ALS works is that it fixes first the user latent factor values and tries to vary the item latent factor values such that the overall SSE reduces. In the next step, the item latent factor values are kept fixed, and the user latent factor values are updated to further reduce the SSE. This keeps alternating between user matrix and item matrix till there can be no more reduction in SSE.

Advantages

1. Content information of the item is not required, and recommendations can be made based on valuable user item interactions.

2. Personalizing experience based on other users.

Limitations

1. Cold Start Problem: If the user has no historical data of item interactions, then the RS cannot predict the K-nearest neighbors for the new user and hence cannot make recommendations.

2. Missing Values: Since the items are huge in numbers and very few users interact with all the items, some items are never rated by users and can't be recommended.

3. Cannot Recommend New or Unrated Items: If an item is new and yet to be seen by the user, it can't be recommended to existing users till other users interact with it.

4. Poor Accuracy: It doesn't perform that well as many components keep changing such as interests of users, limited shelf life of items, and very few ratings of items.

Hybrid Recommender Systems

As the name suggests, hybrid recommender systems include inputs from multiple recommender systems, making it more powerful and relevant in terms of meaningful recommendations to users. As we have seen, there are a few limitations in using the individual RSs, but in combination they overcome few of those and hence are able to recommend items or information that users find more useful and personalized. The hybrid RS can be built in specific ways to suit the requirement of the business. One of the approaches is to build individual RSs and combine the recommendations from multiple RSs before recommending to the user as shown in Figure 8-5.

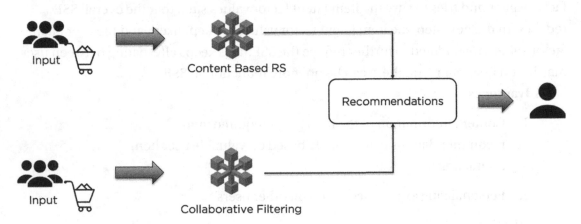

Figure 8-5. *Combining recommendations*

The other approach is leveraging content-based recommender strengths and using them as input for collaborative filtering–based recommendations to provide better recommendations to users. This approach can also be reversed, and collaborative filtering can be used as input for content-based recommendations as shown in Figure 8-6.

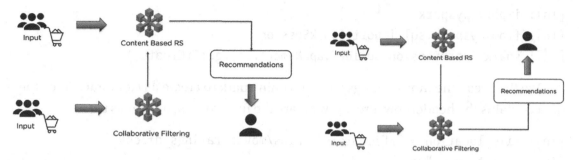

Figure 8-6. *Hybrid recommendations*

Hybrid recommendations also include using other types of recommendations such as demographic-based and knowledge-based recommendations to enhance their performance. Hybrid RSs have become an integral part of various businesses to help their users consume the right content, hence driving a lot of value.

Code

This section of the chapter focuses on building a recommender system using ALS in PySpark and a Databricks notebook.

Note   The complete dataset along with the code is available for reference on the GitHub repo of this book and executes best on Spark 3.1.0 or higher.

Data Info

The dataset that we are going to use for this chapter is a subset from a famous open source movie lens dataset and contains a total of 0.1M records with three columns (User_Id, title, rating). We will train our recommender model using 75% of the data and test it on the rest of the user ratings (25%). We will try to predict the next best movie that the user has not seen yet but might like to watch. We will build a function at the end to recommend "n" number of movies that the user would like based on their history. We start the Databricks cluster and import the movie ratings dataset. We then import SparkSession and create a new SparkSession object to use Spark:

```
[In]: import pyspark
[In]: from pyspark.sql import SparkSession
[In]: spark=SparkSession.builder.appName('rc').getOrCreate()
```

We then read the movie ratings dataset within Spark to create a Dataframe. Since the data contains the header row already, we mark the first_row_is_header as true:

```
[In]: file_location = "/FileStore/tables/movie_ratings_df.csv"
[In]: file_type = "csv"

[In]: infer_schema = "false"
[In]: first_row_is_header = "true"
[In]: delimiter = ","

[In]: df = spark.read.format(file_type) \
      .option("inferSchema", infer_schema) \
      .option("header", first_row_is_header) \
      .option("sep", delimiter) \
      .load(file_location)
```

```
[In]: display(df)
[Out]:
```

	userId	title	rating
1	196	Kolya (1996)	3
2	63	Kolya (1996)	3
3	226	Kolya (1996)	5
4	154	Kolya (1996)	3
5	306	Kolya (1996)	5
6	296	Kolya (1996)	4

As we can see, the Dataframe contains three columns. Let us explore the dataset further to understand the total number of unique users and movies being rated:

```
[In]:print((df.count(), len(df.columns)))
[Out]: (100000,3)
```

So the preceding output confirms the size of our dataset, and we can then validate the datatypes of the input values to check if we need to change/cast any column datatypes:

```
[In]: df.printSchema()
[Out]:
```

```
root
 |-- userId: string (nullable = true)
 |-- title: string (nullable = true)
 |-- rating: string (nullable = true)
```

There are a total of three columns, and one of them seems to be of the string datatype. We will have to convert it into numerical form to build the recommender system. We now view a few rows of the dataframe using the rand function to shuffle the records in random order:

```
[In]: from pyspark.sql.functions import *
[In]: df.orderBy(rand()).show(10,False)
[Out]:
```

```
+------+--------------------------------------+------+
|userId|title                                 |rating|
+------+--------------------------------------+------+
|496   |Sleepless in Seattle (1993)           |1     |
|152   |Gone with the Wind (1939)             |5     |
|116   |Female Perversions (1996)             |3     |
|44    |Amadeus (1984)                        |4     |
|497   |Interview with the Vampire (1994)     |4     |
|776   |Birds, The (1963)                     |3     |
|903   |Grifters, The (1990)                  |4     |
|886   |This Is Spinal Tap (1984)             |4     |
|74    |Jerry Maguire (1996)                  |4     |
|272   |Raising Arizona (1987)                |5     |
+------+--------------------------------------+------+
```

```
[In]:df.groupBy('userId').count().orderBy('count',ascending=False).
show(10,False)
[Out]:
```

```
+------+-----+
|userId|count|
+------+-----+
|405   |737  |
|655   |685  |
|13    |636  |
|450   |540  |
|276   |518  |
|416   |493  |
|537   |490  |
|303   |484  |
|234   |480  |
|393   |448  |
+------+-----+
```

```
[In]: df.groupBy('userId').count().orderBy('count',ascending=True).
show(10,False)
[Out]:
```

```
+------+-----+
|userId|count|
+------+-----+
|732   |20   |
|631   |20   |
|636   |20   |
|926   |20   |
|93    |20   |
|300   |20   |
|572   |20   |
|596   |20   |
|685   |20   |
|34    |20   |
+------+-----+
```

The user with the highest number of records has rated 737 movies, and each user has rated at least 20 movies:

```
[In]: df.groupBy('title').count().orderBy('count',ascending=False).
show(10,False)
[Out]:
```

```
+-------------------------------+-----+
|title                          |count|
+-------------------------------+-----+
|Star Wars (1977)               |583  |
|Contact (1997)                 |509  |
|Fargo (1996)                   |508  |
|Return of the Jedi (1983)      |507  |
|Liar Liar (1997)               |485  |
|English Patient, The (1996)    |481  |
|Scream (1996)                  |478  |
|Toy Story (1995)               |452  |
|Air Force One (1997)           |431  |
|Independence Day (ID4) (1996)  |429  |
+-------------------------------+-----+
```

The movie with the highest number of ratings is *Star Wars* (1977) and has been rated 583 times, and each movie has been rated by at least one user. We now cast the datatype for the title column. We convert the movie title column from categorical to numerical values using StringIndexer. We import StringIndexer and IndexToString from the PySpark library:

```
[In]: from pyspark.sql.functions import *
[In]: from pyspark.ml.feature import StringIndexer,IndexToString
[In]: from pyspark.sql.types import DoubleType
[In]: df = df.withColumn("userId", df.userId.cast(DoubleType()))
[In]: df = df.withColumn("rating", df.rating.cast(DoubleType()))
```

Next, we create the StringIndexer object by mentioning the input column and output column. Then we fit the object on the dataframe and apply it on the movie title column to create a new dataframe with numerical values:

```
[In]: stringIndexer = StringIndexer(inputCol="title",
outputCol="title_new")

[In]: model = stringIndexer.fit(df)

[In]: indexed = model.transform(df)
```

Let's validate the numerical values of the title column by viewing a few rows from the new dataframe (indexed):

```
[In]: indexed.show(10)
[Out]:
```

```
+-------+------------+------+---------+
|userId|       title|rating|title_new|
+-------+------------+------+---------+
| 196.0|Kolya (1996)|   3.0|    287.0|
|  63.0|Kolya (1996)|   3.0|    287.0|
| 226.0|Kolya (1996)|   5.0|    287.0|
| 154.0|Kolya (1996)|   3.0|    287.0|
| 306.0|Kolya (1996)|   5.0|    287.0|
| 296.0|Kolya (1996)|   4.0|    287.0|
|  34.0|Kolya (1996)|   5.0|    287.0|
| 271.0|Kolya (1996)|   4.0|    287.0|
| 201.0|Kolya (1996)|   4.0|    287.0|
| 209.0|Kolya (1996)|   4.0|    287.0|
+-------+------------+------+---------+
```

As we can see, now we have an additional column (title_new) with numerical values representing the movie titles. Just to validate the movie counts, we rerun the groupBy function on the new dataframe:

```
[In]: indexed.groupBy('title_new').count().orderBy('count',ascending=
False).show(10,False)
[Out]:
```

```
+---------+-----+
|title_new|count|
+---------+-----+
|0.0      |583  |
|1.0      |509  |
|2.0      |508  |
|3.0      |507  |
|4.0      |485  |
|5.0      |481  |
|6.0      |478  |
|7.0      |452  |
|8.0      |431  |
|9.0      |429  |
+---------+-----+
```

Now that we have prepared the data for building the recommender model, we can split the dataset into training and test sets. We split it in 75/25 ratio to train the model and test its accuracy:

```
[In]: train,test=indexed.randomSplit([0.75,0.25])
```

```
[In]: train.count()
[Out]: 75159
```

```
[In]: test.count()
[Out]: 24841
```

We import the ALS function from PySpark's ML library and build the model on the training set. There are multiple hyperparameters that can be tuned to improve the performance of the model. Two of the important ones are as follows: nonnegative ='True' doesn't create negative ratings in recommendations, and coldStartStrategy='drop' prevents any NaN rating predictions:

```
[In]: from pyspark.ml.recommendation import ALS
[In]:rec=ALS(maxIter=10,regParam=0.01,userCol='userId',itemCol='title_new',
ratingCol='rating',nonnegative=True,coldStartStrategy="drop")
```

```
[In]: rec_model=rec.fit(train)
```

The final part of the entire exercise is to check the performance of the model on unseen or test data. We use the transform function to make predictions on the test data and RegressionEvaluate to check the RMSE value of the model on test data:

```
[In]: predicted_ratings=rec_model.transform(test)
```

```
[In]: predicted_ratings.printSchema()
root
 |-- userId: integer (nullable = true)
 |-- title: string (nullable = true)
 |-- rating: integer (nullable = true)
 |-- title_new: double (nullable = false)
 |-- prediction: float (nullable = false)
```

```
[In]: predicted_ratings.orderBy(rand()).show(10)
[Out]:
```

```
+-------+--------------------+------+---------+----------+
|userId|               title|rating|title_new|prediction|
+-------+--------------------+------+---------+----------+
| 151.0|Fried Green Tomat...|   4.0|    197.0| 3.5980523|
|  15.0|Island of Dr. Mor...|   1.0|    562.0| 1.8102996|
| 821.0|      Top Gun (1986)|   4.0|     94.0| 3.9421341|
|  11.0|        Brazil (1985)|  3.0|    109.0| 3.0553205|
| 623.0|Four Weddings and...|   4.0|     66.0| 3.6468706|
| 450.0|  Being There (1979)|   4.0|    290.0| 4.3743834|
| 346.0|Batman Forever (1...|   4.0|    297.0| 2.5176854|
| 226.0|   Apollo 13 (1995)|   4.0|     51.0| 4.1644554|
| 151.0| Citizen Kane (1941)|   4.0|    119.0|  4.786707|
| 301.0|        Speed (1994)|   4.0|     84.0| 3.4523523|
+-------+--------------------+------+---------+----------+
```

```
[In]: from pyspark.ml.evaluation import RegressionEvaluator
```

```
[In]: evaluator=RegressionEvaluator(metricName='rmse',predictionCol=
'prediction',labelCol='rating')
```

```
[In]: rmse=evaluator.evaluate(predictions)
```

```
[In] : print(rmse)
```

```
[Out]: 1.01840
```

The RMSE is not very high. We are making an error of one point in actual rating
and predicted rating. This can be improved further by tuning model parameters and
using the hybrid approach. After checking the performance of the model and tuning
the hyperparameters, we can move ahead to recommend top movies to users, which
they have not seen and might like. The first step is to create a list of unique movies in the
dataframe:

```
[In]: unique_movies=indexed.select('title_new').distinct()
```

```
[In]: unique_movies.count()
```

```
[Out]: 1664
```

So we have in total 1664 distinct movies in the dataframe

```
[In]: a = unique_movies.alias('a')
```

We can select any user within the dataset for which we need to recommend other movies. In our case, we go ahead with userId = 85.0.

```
[In]: user_id=85.0
```

We will filter the movies that this active user has already rated or seen:

```
[In]: watched_movies=indexed.filter(indexed['userId'] == user_id).
select('title_new').distinct()
```

```
[In]: watched_movies.count()
[Out]: 287
```

```
[In]: b=watched_movies.alias('b')
```

So there are a total of 287 unique movies out of 1664 movies that this active user has already rated. So we would want to recommend movies from the remaining 1377 items. We now combine both the tables to find the movies that we can recommend by filtering null values from the joined table:

```
[In]: total_movies = a.join(b, a.title_new == b.title_new,how='left')
```

```
[In]: total_movies.show(10,False)
```

```
[Out]:
```

```
+---------+---------+
|title_new|title_new|
+---------+---------+
|305.0    |305.0    |
|596.0    |null     |
|299.0    |null     |
|769.0    |null     |
|692.0    |null     |
|934.0    |null     |
|1051.0   |null     |
|496.0    |null     |
|558.0    |558.0    |
|170.0    |null     |
+---------+---------+
```

```
[In]: remaining_movies=total_movies.where(col("b.title_new").isNull()).
select(a.title_new).distinct()

[In]: remaining_movies.count()

[Out]: 1377

[In]: remaining_movies=remaining_movies.withColumn("userId",lit(int(
user_id)))
[In]: remaining_movies.show(10,False)

[Out]:
```

```
+---------+------+
|title_new|userId|
+---------+------+
|596.0    |85    |
|299.0    |85    |
|769.0    |85    |
|692.0    |85    |
|934.0    |85    |
|1051.0   |85    |
|496.0    |85    |
|170.0    |85    |
|184.0    |85    |
|576.0    |85    |
+---------+------+
```

Finally, we can now make the predictions on this remaining movies dataset for the active user using the recommender model that we built earlier. We filter only the top few recommendations that have the highest predicted ratings:

```
[In]: recommendations=rec_model.transform(remaining_movies).orderBy
('prediction',ascending=False)

[In]: recommendations.show(5,False)
[Out]:
```

```
+---------+------+----------+
|title_new|userId|prediction|
+---------+------+----------+
|1277.0   |85    |5.206085  |
|1054.0   |85    |4.998125  |
|963.0    |85    |4.8784885 |
|1288.0   |85    |4.735536  |
|1155.0   |85    |4.7105002 |
+---------+------+----------+
```

So movie titles 1277 and 1054 have the highest predicted rating for this active user (85). We can make it more intuitive by adding the movie title back to the recommendations. We use the IndexToString function to create an additional column that returns the movie title:

```
[In]:
movie_title = IndexToString(inputCol="title_new",
outputCol="title",labels=model.labels)
[In]:final_recommendations=movie_title.transform(recommendations)

[In]: final_recommendations.show(10,False)
```

[Out]:

```
+---------+------+----------+---------------------------------------------------------------------+
|title_new|userId|prediction|title                                                                |
+---------+------+----------+---------------------------------------------------------------------+
|1277.0   |85    |5.206085  |Mina Tannenbaum (1994)                                               |
|1054.0   |85    |4.998125  |Primary Colors (1998)                                               |
|963.0    |85    |4.8784885 |Inspector General, The (1949)                                       |
|1288.0   |85    |4.735536  |Whole Wide World, The (1996)                                        |
|1155.0   |85    |4.7105002 |Horseman on the Roof, The (Hussard sur le toit, Le) (1995)          |
|921.0    |85    |4.6698713 |For Whom the Bell Tolls (1943)                                      |
|1128.0   |85    |4.618379  |Incognito (1997)                                                    |
|632.0    |85    |4.617119  |Great Dictator, The (1940)                                          |
|892.0    |85    |4.6028943 |Double vie de Véronique, La (Double Life of Veronique, The) (1991)  |
|1165.0   |85    |4.566659  |Murder, My Sweet (1944)                                             |
+---------+------+----------+---------------------------------------------------------------------+
```

So the recommendations for the userId (85) are *Mina Tannenbaum* (1994) and *Primary Colors* (1998). This can be nicely wrapped in a single function that executes the preceding steps in sequence and generates recommendations for the active user as shown in the following:

```
#create function to recommend top 'n' movies to any particular user
def top_movies(user_id,n):
    """

    This function returns the top 'n' movies that user has not seen yet but
    might like

    """
    #assigning alias name 'a' to unique movies df
    a = unique_movies.alias('a')

    #creating another dataframe which contains already watched movie by
    active user
    watched_movies=indexed.filter(indexed['userId'] == user_id).
    select('title_new')

    #assigning alias name 'b' to watched movies df
    b=watched_movies.alias('b')

    #joining both tables on left join
    total_movies = a.join(b, a.title_new == b.title_new,how='left')

    #selecting movies which active user is yet to rate or watch
    remaining_movies=total_movies.where(col("b.title_new").isNull()).
    select(a.title_new).distinct()

    #adding new column of user_Id of active useer to remaining movies df
    remaining_movies=remaining_movies.withColumn("userId",lit(int(
    user_id)))

    #making recommendations using ALS recommender model and selecting only
    top 'n' movies
    recommendations=rec_model.transform(remaining_movies).orderBy('
    prediction',ascending=False).limit(n)

    #adding columns of movie titles in recommendations
    movie_title = IndexToString(inputCol="title_new",
    outputCol="title",labels=model.labels)
    final_recommendations=movie_title.transform(recommendations)
    output=final_recommendations.select('userId','title_new','title')
```

```
    #return the recommendations to active user
    return output.show(n,False)

[In]: top_movies(85.0,10)
[Out]:

+------+---------+--------------------------------------+
|userId|title_new|title                                 |
+------+---------+--------------------------------------+
|67    |1040.0   |Boys of St. Vincent, The (1993)       |
|67    |1495.0   |Love and Death on Long Island (1997)|
|67    |1281.0   |Promesse, La (1996)                   |
|67    |1028.0   |Braindead (1992)                      |
|67    |1517.0   |Slingshot, The (1993)                 |
|67    |845.0    |Pillow Book, The (1995)               |
|67    |831.0    |Pollyanna (1960)                      |
|67    |840.0    |Friday (1995)                         |
|67    |1122.0   |Faithful (1996)                       |
|67    |1024.0   |Amityville II: The Possession (1982)|
+------+---------+--------------------------------------+
```

Conclusion

In this chapter, we went over various types of recommendation models along with
strengths and limitations of each. We then created a collaborative filtering–based
recommender system in PySpark using the ALS method to recommend movies to users.

CHAPTER 9

Natural Language Processing

This is the last chapter of the book and focuses on the techniques to tackle text data using PySpark. Today text-form data is being generated at a lightning pace with multiple social media platforms offering users the options to share their reviews, suggestions, comments, etc. The area that focuses on making machines learn and understand textual data to perform some useful tasks is known as Natural Language Processing. Text data could be structured or unstructured, and we must apply multiple steps to make it analysis ready. The NLP field is already a huge area of research and has an immense number of applications being developed that use text data such as chatbots, speech recognition, language translation, recommender systems, spam detection, and sentiment analysis. This chapter demonstrates a series of steps to process text data and apply Machine Learning algorithms on it. It also showcases sequence embeddings that are learned using word2vec in PySpark as a bonus part.

Steps Involved in NLP

Text data can be very messy sometimes, and it needs careful attention to bring it to a stage where it can be used in the right way. There are multiple ways in which text data can be cleaned and refined. For example, regular expressions are very powerful when it comes to filtering, cleaning, and standardizing text data. However, regular expressions are not the focus area in this chapter. Rather, we look at the steps to prepare text data in a form where we can fit a ML model on it. The five major steps involved in handling text data for ML modeling are

1. Reading the corpus

2. Tokenization

© Pramod Singh 2022
P. Singh, *Machine Learning with PySpark*, https://doi.org/10.1007/978-1-4842-7777-5_9

3. Cleaning/stopword removal

4. Stemming

5. Converting into numerical form

Before jumping into the steps to load and clean text data, let's get familiar with a term known as *corpus* as this would keep appearing in the rest of the chapter.

Corpus

A corpus is known as an entire collection of text documents, for example, a collection of emails, messages, or user reviews. This group of individual text items is known as a corpus. The next step in text processing is tokenization

Tokenize

The method of dividing the given text sequence/sentence or collection of words of a text document into individual words is known as tokenization. It removes the unnecessary characters such as punctuations. The final units post-tokenization are known as tokens. Let's say we have the following text:

Input: He really liked the London City. He is there for two more days.

Tokenization would result in the following tokens. We end up with 13 tokens for the input text:

He, really, liked, the, London, City, He, is, there, for, two, more, days

Let us see how we can do tokenization using PySpark. The first step is to create a SparkSession object to use Spark. We take some sample user IDs and reviews to create a text Dataframe as shown in the following:

```
[In]: import pyspark
[In]: from pyspark.sql import SparkSession
[In]: spark=SparkSession.builder.appName('nlp').getOrCreate()

[In]: df=spark.createDataFrame([(1,'I really liked this movie'),
                      (2,'I would recommend this movie to my friends'),
                      (3,'movie was alright but acting was horrible'),
                      (4,'I am never watching that movie ever again')],
                     ['user_id','review'])
```

```
[In]: df.show(4,False)
[Out]:
```

```
+-------+----------------------------------------+
|user_id|review                                  |
+-------+----------------------------------------+
|1      |I really liked this movie               |
|2      |I would recommend this movie to my friends|
|3      |movie was alright but acting was horrible |
|4      |I am never watching that movie ever again |
+-------+----------------------------------------+
```

In this Dataframe, we have four sentences for tokenization. The next step is to import Tokenizer from the Spark library. We must then pass the input column and name the output column after tokenization. We use the transform function to apply tokenization to the review column:

```
[In]: from pyspark.ml.feature import Tokenizer
[In]: tokenization=Tokenizer(inputCol='review',outputCol='tokens')
[In]: tokenized_df=tokenization.transform(df)
[In]: tokenized_df.show(4,False)
[Out]:
```

```
+-------+----------------------------------------+------------------------------------------------+
|user_id|review                                  |tokens                                          |
+-------+----------------------------------------+------------------------------------------------+
|1      |I really liked this movie               |[i, really, liked, this, movie]                 |
|2      |I would recommend this movie to my friends|[i, would, recommend, this, movie, to, my, friends]|
|3      |movie was alright but acting was horrible |[movie, was, alright, but, acting, was, horrible] |
|4      |I am never watching that movie ever again |[i, am, never, watching, that, movie, ever, again] |
+-------+----------------------------------------+------------------------------------------------+
```

We get a new column named tokens, which contains the tokens for each sentence.

Stopword Removal

As you can observe, the tokens column contains very common words such as "this," "the," "to," "was," "that," etc. These words are known as stopwords, and they seem to add very little value to the analysis. If they are to be used in analysis, it increases the computation overheads without adding too much signal. Hence, it's preferred to drop these stopwords from the tokens. In PySpark, we use StopWordsRemover to remove the stopwords:

```
[In]:from pyspark.ml.feature import StopWordsRemover
[In]: stopword_removal=StopWordsRemover(inputCol='tokens',outputCol='refined_
tokens')
```

We then pass the tokens as the input column and name the output column as refined tokens.

```
[In]: refined_df=stopword_removal.transform(tokenized_df)
[In]: refined_df.select(['user_id','tokens','refined_tokens']).
show(4,False)
[Out]:
```

```
+-------+---------------------------------------------------+----------------------------------+
|user_id|tokens                                             |refined_tokens                    |
+-------+---------------------------------------------------+----------------------------------+
|1      |[i, really, liked, this, movie]                    |[really, liked, movie]            |
|2      |[i, would, recommend, this, movie, to, my, friends]|[recommend, movie, friends]       |
|3      |[movie, was, alright, but, acting, was, horrible]  |[movie, alright, acting, horrible]|
|4      |[i, am, never, watching, that, movie, ever, again] |[never, watching, movie, ever]    |
+-------+---------------------------------------------------+----------------------------------+
```

As you can observe, the stopwords like "I," "this," "was," "am," "but," and "that" are removed from the tokens column.

Bag of Words

Now that we have the key tokens created from the text data, we now need a mechanism to convert the tokens into a numerical form because we know for a fact that a Machine Learning algorithm works on numerical data. Text data is generally unstructured and varies in its length. Bag of words allows to convert the text data into numerical vector form by considering the occurrence of the words in text documents, for example:

Doc 1: The best thing in life is to travel

Doc 2: Travel is the best medicine

Doc 3: One should travel more often

The list of unique words appearing in all the documents is known as *vocabulary*. In the preceding example, we have a total of 13 unique words appearing that are part of the vocab. Any of these three documents can be represented by this vector of fixed size 13 using a Boolean value (1 or 0):

The	Best	thing	in	life	is	to	travel	medicine	one	should	more	often

Doc 1:

The	Best	thing	in	life	is	to	travel	medicine	one	should	more	often
1	1	1	1	1	1	1	1	0	0	0	0	0

Doc 2:

The	Best	thing	in	life	is	to	travel	medicine	one	should	more	often
1	1	0	0	0	1	0	1	1	0	0	0	0

Doc 3:

The	Best	thing	in	life	is	to	travel	medicine	one	should	more	often
0	0	0	0	0	0	0	1	0	1	1	1	1

Bag of words does not consider the order of words in the document and the semantic meaning of the word and hence is the most baseline method to represent the text data in numerical form. There are other ways by which we can convert the textual data into numerical form, which are covered in the following section. We will use PySpark to go through each one of these methods.

CountVectorizer

In bag of words, we saw the representation of the occurrence of a word by simply 1 or 0 and did not consider the frequency of the word. The count vectorizer instead takes the total count of the tokens appearing in the document. We will use the same text documents that we created earlier during tokenization. We first import CountVectorizer:

```
[In]: from pyspark.ml.feature import CountVectorizer
[In]: count_vec=CountVectorizer(inputCol='refined_tokens',
outputCol='features')
[In]: cv_df=count_vec.fit(refined_df).transform(refined_df)
[In]: cv_df.select(['user_id','refined_tokens','features']).show(4,False)
[Out]:
```

```
+-------+----------------------------------+------------------------------------+
|user_id|refined_tokens                    |features                            |
+-------+----------------------------------+------------------------------------+
|1      |[really, liked, movie]            |(11,[0,3,6],[1.0,1.0,1.0])          |
|2      |[recommend, movie, friends]       |(11,[0,1,7],[1.0,1.0,1.0])          |
|3      |[movie, alright, acting, horrible]|(11,[0,8,9,10],[1.0,1.0,1.0,1.0])|
|4      |[never, watching, movie, ever]    |(11,[0,2,4,5],[1.0,1.0,1.0,1.0]) |
+-------+----------------------------------+------------------------------------+
```

As we can observe, each row is represented as a dense vector. It shows that the vector length is 11 and the first sentence contains three values at the 0th, 4th, and 9th indexes.

To validate the vocabulary of the count vectorizer, we can simply use the vocabulary function:

```
[In]: count_vec.fit(refined_df).vocabulary
[Out]:
```

```
                           ['movie',
                            'acting',
                            'friends',
                            'liked',
                            'never',
                            'horrible',
                            'alright',
                            'recommend',
                            'really',
                            'watching',
                            'ever']
```

Hence, the vocabulary size for the preceding sentences is 11; and if you look at the features carefully, they are like the input feature vector that we have been using for Machine Learning in PySpark. The drawback of using the CountVectorizer method is that it doesn't consider the co-occurrences of words in other documents. In simple terms, the words appearing often would have larger impact on the feature vector.

Hence, another approach to convert text data into numerical form is TF-IDF (Term Frequency – Inverse Document Frequency).

TF-IDF

This method tries to normalize the frequency of token occurrence based on other documents. The whole idea is to give more weight to the token if appearing high number of times in the same document but penalize if it is appearing higher number of times in other documents as well. This indicates that the token is common across the corpus and is not as important as its frequency in the current document indicates.

Term Frequency: Score based on the frequency of the word in the current document

Inverse Document Frequency: Score based on the number of documents that contain the current word

Now, we create features based on TF-IDF in PySpark using the same refined df Dataframe:

```
[In]: from pyspark.ml.feature import HashingTF,IDF
[In]: hashing_vec=HashingTF(inputCol='refined_tokens',outputCol='tf_features')
[In]: hashing_df=hashing_vec.transform(refined_df)
[In]: hashing_df.select(['user_id','refined_tokens','tf_features']).
show(4,False)
[Out]:
```

```
+-------+----------------------------------+------------------------------------------------------+
|user_id|refined_tokens                    |tf_features                                           |
+-------+----------------------------------+------------------------------------------------------+
|1      |[really, liked, movie]            |(262144,[99172,210223,229264],[1.0,1.0,1.0])          |
|2      |[recommend, movie, friends]       |(262144,[68228,130047,210223],[1.0,1.0,1.0])          |
|3      |[movie, alright, acting, horrible]|(262144,[95685,171118,210223,236263],[1.0,1.0,1.0,1.0])|
|4      |[never, watching, movie, ever]    |(262144,[63139,113673,203802,210223],[1.0,1.0,1.0,1.0])|
+-------+----------------------------------+------------------------------------------------------+
```

```
[In]: tf_idf_vec=IDF(inputCol='tf_features',outputCol='tf_idf_features')
[In]: tf_idf_df=tf_idf_vec.fit(hashing_df).transform(hashing_df)
[In]: tf_idf_df.select(['user_id','tf_idf_features']).show(4,False)
[Out]:
```

```
+-------+----------------------------------------------------------------------------------------------+
|user_id|tf_idf_features                                                                               |
+-------+----------------------------------------------------------------------------------------------+
|1      |(262144,[99172,210223,229264],[0.9162907318741551,0.0,0.9162907318741551])                    |
|2      |(262144,[68228,130047,210223],[0.9162907318741551,0.9162907318741551,0.0])                    |
|3      |(262144,[95685,171118,210223,236263],[0.9162907318741551,0.9162907318741551,0.0,0.9162907318741551])|
|4      |(262144,[63139,113673,203802,210223],[0.9162907318741551,0.9162907318741551,0.9162907318741551,0.0])|
+-------+----------------------------------------------------------------------------------------------+
```

Text Classification Using Machine Learning

Now that we understand the steps involved in dealing with text processing and feature vectorization, we can build a text classification model and use it for predictions on text data. The dataset that we are going to use is a sample of the open source movie lens reviews data, and we're going to predict the sentiment of the given review (positive or negative). Let's start with reading the text data first and creating a Spark Dataframe:

```
[In]: file_location = "/FileStore/tables/Movie_reviews.csv"
[In]: file_type = "csv"

[In]: infer_schema = "false"
[In]: first_row_is_header = "true"
[In]: delimiter = ","

[In]: text_df = spark.read.format(file_type) \
        .option("inferSchema", infer_schema) \
        .option("header", first_row_is_header) \
        .option("sep", delimiter) \
        .load(file_location)

[In]: display(text_df)
[Out]:
```

	Review	Sentiment
1	The Da Vinci Code book is just awesome.	1
2	this was the first clive cussler i've ever read, but even books like Rel	1
3	i liked the Da Vinci Code a lot.	1
4	i liked the Da Vinci Code a lot.	1
5	I liked the Da Vinci Code but it ultimatly didn't seem to hold it's own.	1
6	that's not even an exaggeration) and at midnight we went to Wal-Mart to	1

```
[In]: text_df.printSchema()
[Out]:

        root
         |-- Review: string (nullable = true)
         |-- Sentiment: string (nullable = true)
```

As we can see, the Sentiment column is StringType, and we will need to convert it into an integer or float type going forward:

```
[In]: text_df.count()
[Out]: 7087
```

We have close to 7k records, out of which some might not be labeled properly. Hence, we filter only those records that are labeled correctly:

```
[In]: text_df=text_df.filter(((text_df.Sentiment =='1') |
(text_df.Sentiment =='0')))
[In]: text_df.count()
[Out]: 6990
```

Some of the records got filtered out, and we are now left with 6990 records for the analysis. The next step is to validate the number of reviews for each class:

```
[In]: text_df.groupBy('Sentiment').count().show()
[Out]:
+---------+-----+
|Sentiment|count|
+---------+-----+
|        0| 3081|
|        1| 3909|
+---------+-----+
```

We are dealing with a balanced dataset here as both classes have an almost similar number of reviews. Let us look at a few of the records in the dataset. As a next step, we create a new integer-type Label column and drop the original Sentiment column, which was string type:

```
[In]:text_df=text_df.withColumn("Label",text_df.Sentiment.cast('float')).
drop('Sentiment')

[In]: from pyspark.sql.functions import rand
[In]: text_df.orderBy(rand()).show(10,False)
[Out]:
```

```
+----------------------------------------------------------------------+------+
|Review                                                                |Label|
+----------------------------------------------------------------------+------+
|Harry Potter is AWESOME I don't care if anyone says differently!..    |1.0  |
|I love Harry Potter..                                                 |1.0  |
|As I sit here, watching the MTV Movie Awards, I am reminded of how much|0.0  |
|Oh, and Brokeback Mountain is a TERRIBLE movie...                     |0.0  |
|Which is why i said silent hill turned into reality coz i was hella like|1.0  |
|Brokeback Mountain was so awesome.                                    |1.0  |
|I love The Da Vinci Code...                                           |1.0  |
|i love being a sentry for mission impossible and a station for bonkers.|1.0  |
|I thought Brokeback Mountain was an awful movie.                      |0.0  |
|friday hung out with kelsie and we went and saw The Da Vinci Code SUCKED|0.0  |
+----------------------------------------------------------------------+------+
```

We also include an additional column that captures the length of the review:

```
[In]: from pyspark.sql.functions import length
[In]: text_df=text_df.withColumn('length',length(text_df['Review']))
[In]: text_df.orderBy(rand()).show(10,False)
[Out]:
```

```
+----------------------------------------------------------------------+-----+------+
|Review                                                                |Label|length|
+----------------------------------------------------------------------+-----+------+
|dudeee i LOVED brokeback mountain!!!!                                 |1.0  |37    |
|The Da Vinci Code was awesome, I can't wait to read it...             |1.0  |57    |
|Brokeback mountain was beautiful...                                   |1.0  |35    |
|Then snuck into Brokeback Mountain, which is the most depressing movie I|0.0  |72    |
|The Da Vinci Code was awesome, I can't wait to read it...             |1.0  |57    |
|I like Mission Impossible movies because you never know who's on the rig|1.0  |72    |
|children's texts-fantasy perhaps most obviously-is often criticized for |1.0  |71    |
|i love kirsten / leah / kate escapades and mission impossible tom as wel|1.0  |72    |
|Oh oh oh and I loved The Da Vinci Code!                               |1.0  |39    |
|As sad as it may be that I am now dating my sister, I enjoyed our time t|1.0  |72    |
+----------------------------------------------------------------------+-----+------+
```

```
[In]: text_df.groupBy('Label').agg({'Length':'mean'}).show()
[Out]:
```

```
+-----+-----------------+
|Label|      avg(Length)|
+-----+-----------------+
|  1.0|47.61882834484523|
|  0.0|50.95845504706264|
+-----+-----------------+
```

There is no major difference between the average length of the positive review and that of the negative review. The next step is to start the tokenization process and remove stopwords:

```
[In]: tokenization=Tokenizer(inputCol='Review',outputCol='tokens')
[In]: tokenized_df=tokenization.transform(text_df)
[In]: tokenized_df.show()
```

```
+--------------------+-----+------+--------------------+
|              Review|Label|length|              tokens|
+--------------------+-----+------+--------------------+
|The Da Vinci Code...|  1.0|    39|[the, da, vinci, ...|
|this was the firs...|  1.0|    72|[this, was, the, ...|
|i liked the Da Vi...|  1.0|    32|[i, liked, the, d...|
|i liked the Da Vi...|  1.0|    32|[i, liked, the, d...|
|I liked the Da Vi...|  1.0|    72|[i, liked, the, d...|
|that's not even a...|  1.0|    72|[that's, not, eve...|
|I loved the Da Vi...|  1.0|    72|[i, loved, the, d...|
|i thought da vinc...|  1.0|    57|[i, thought, da, ...|
|The Da Vinci Code...|  1.0|    45|[the, da, vinci, ...|
|I thought the Da ...|  1.0|    51|[i, thought, the,...|
|The Da Vinci Code...|  1.0|    68|[the, da, vinci, ...|
|The Da Vinci Code...|  1.0|    62|[the, da, vinci, ...|
|then I turn on th...|  1.0|    66|[then, i, turn, o...|
|The Da Vinci Code...|  1.0|    34|[the, da, vinci, ...|
|i love da vinci c...|  1.0|    24|[i, love, da, vin...|
|i loved da vinci ...|  1.0|    23|[i, loved, da, vi...|
|TO NIGHT:: THE DA...|  1.0|    52|[to, night::, the...|
```

```
[In]:stopword_removal=StopWordsRemover(inputCol='tokens',outputCol=
'refined_tokens')
[In]: refined_text_df=stopword_removal.transform(tokenized_df)
[In]: refined_text_df.show()
[Out]:
```

```
+--------------------+-----+------+--------------------+--------------------+
|              Review|Label|length|              tokens|     refined_tokens|
+--------------------+-----+------+--------------------+--------------------+
|The Da Vinci Code...|  1.0|    39|[the, da, vinci, ...|[da, vinci, code,...|
|this was the firs...|  1.0|    72|[this, was, the, ...|[first, clive, cu...|
|i liked the Da Vi...|  1.0|    32|[i, liked, the, d...|[liked, da, vinci...|
|i liked the Da Vi...|  1.0|    32|[i, liked, the, d...|[liked, da, vinci...|
|I liked the Da Vi...|  1.0|    72|[i, liked, the, d...|[liked, da, vinci...|
|that's not even a...|  1.0|    72|[that's, not, eve...|[even, exaggerati...|
|I loved the Da Vi...|  1.0|    72|[i, loved, the, d...|[loved, da, vinci...|
|i thought da vinc...|  1.0|    57|[i, thought, da, ...|[thought, da, vin...|
|The Da Vinci Code...|  1.0|    45|[the, da, vinci, ...|[da, vinci, code,...|
|I thought the Da ...|  1.0|    51|[i, thought, the,...|[thought, da, vin...|
|The Da Vinci Code...|  1.0|    68|[the, da, vinci, ...|[da, vinci, code,...|
|The Da Vinci Code...|  1.0|    62|[the, da, vinci, ...|[da, vinci, code,...|
|then I turn on th...|  1.0|    66|[then, i, turn, o...|[turn, light, rad...|
|The Da Vinci Code...|  1.0|    34|[the, da, vinci, ...|[da, vinci, code,...|
|i love da vinci c...|  1.0|    24|[i, love, da, vin...|[love, da, vinci,...|
```

Since we're now dealing with tokens only instead of an entire review, it would make more sense to capture the number of tokens in each review rather than using the length of the review. We create another column (token count) that gives the number of tokens in each row:

```
[In]: from pyspark.sql.functions import udf
[In]: from pyspark.sql.types import IntegerType
[In]: from pyspark.sql.functions import *
[In]: len_udf = udf(lambda s: len(s), IntegerType())
[In]: refined_text_df = refined_text_df.withColumn("token_count",
len_udf(col('refined_tokens')))
[In]: refined_text_df.orderBy(rand()).show(10)
[Out]:
```

```
+--------------------+-----+------+--------------------+--------------------+-----------+
|              Review|Label|length|              tokens|     refined_tokens|token_count|
+--------------------+-----+------+--------------------+--------------------+-----------+
|I, too, like Harr...|  1.0|    27|[i,, too,, like, ...|[i,, too,, like, ...|          5|
|"Anyway, thats wh...|  1.0|    49|["anyway,, thats,...|["anyway,, thats,...|          6|
|The first action ...|  1.0|    71|[the, first, acti...|[first, action, t...|          7|
|mission impossibl...|  1.0|    35|[mission, impossi...|[mission, impossi...|          4|
|Da Vinci Code sucks.|  0.0|    20|[da, vinci, code,...|[da, vinci, code,...|          4|
|Da Vinci Code sucks.|  0.0|    20|[da, vinci, code,...|[da, vinci, code,...|          4|
|I love Brokeback ...|  1.0|    29|[i, love, brokeba...|[love, brokeback,...|          3|
|man i loved broke...|  1.0|    31|[man, i, loved, b...|[man, loved, brok...|          4|
|Then snuck into B...|  0.0|    72|[then, snuck, int...|[snuck, brokeback...|          5|
|Brokeback Mountai...|  1.0|    40|[brokeback, mount...|[brokeback, mount...|          4|
+--------------------+-----+------+--------------------+--------------------+-----------+
```

Now that we have the refined tokens after stopword removal, we can use any of the preceding approaches to convert text into numerical features. In this case, we use CountVectorizer for feature vectorization for the Machine Learning model:

```
[In]:count_vec=CountVectorizer(inputCol='refined_
tokens',outputCol='features')
[In]: cv_text_df=count_vec.fit(refined_text_df).transform(refined_text_df)
[In]: cv_text_df.select(['refined_tokens','token_
count','features','Label']).show(10)
[Out]:
```

```
+--------------------+-----------+--------------------+-----+
|      refined_tokens|token_count|            features|Label|
+--------------------+-----------+--------------------+-----+
|[da, vinci, code,...|          5|(2302,[0,1,4,43,2...|  1.0|
|[first, clive, cu...|          9|(2302,[11,51,229,...|  1.0|
|[liked, da, vinci...|          5|(2302,[0,1,4,53,3...|  1.0|
|[liked, da, vinci...|          5|(2302,[0,1,4,53,3...|  1.0|
|[liked, da, vinci...|          8|(2302,[0,1,4,53,6...|  1.0|
|[even, exaggerati...|          6|(2302,[46,229,271...|  1.0|
|[loved, da, vinci...|          8|(2302,[0,1,22,30,...|  1.0|
|[thought, da, vin...|          7|(2302,[0,1,4,228,...|  1.0|
|[da, vinci, code,...|          6|(2302,[0,1,4,33,2...|  1.0|
|[thought, da, vin...|          7|(2302,[0,1,4,223,...|  1.0|
+--------------------+-----------+--------------------+-----+
```

```
[In]: model_text_df=cv_text_df.select(['features','token_count','Label'])
```

Once we have the feature vector for each row, we can make use of VectorAssembler to create input features for the Machine Learning model:

```
[In]: from pyspark.ml.feature import VectorAssembler
[In]: df_assembler = VectorAssembler(inputCols=['features','token_
count'],outputCol='features_vec')
[In]: model_text_df = df_assembler.transform(model_text_df)
[In]: model_text_df.printSchema()

[Out]:

        root
          |-- features: vector (nullable = true)
          |-- token_count: integer (nullable = true)
          |-- Label: float (nullable = true)
          |-- features_vec: vector (nullable = true)
```

We can use any of the classification model on this data, but we proceed with training a logistic regression model:

```
[In]: from pyspark.ml.classification import LogisticRegression
[In]: training_df,test_df=model_text_df.randomSplit([0.75,0.25])
```

To verify the presence of enough records for both classes in train and test sets, we can apply the groupBy function on the Label column:

```
[In]: training_df.groupBy('Label').count().show()
[Out]:

                        +-----+-----+
                        |Label|count|
                        +-----+-----+
                        |  1.0| 2943|
                        |  0.0| 2277|
                        +-----+-----+
```

```
[In]: test_df.groupBy('Label').count().show()
[Out]:
```

```
+-----+-----+
|Label|count|
+-----+-----+
|  1.0|  929|
|  0.0|  779|
+-----+-----+
```

```
[In]: log_reg=LogisticRegression(featuresCol='features_
vec',labelCol='Label').fit(training_df)
[In]: results=log_reg.evaluate(test_df).predictions
[In]: results.show()
```

```
+--------------------+-----------+-----+--------------------+--------------------+--------------------+----------+
|            features|token_count|Label|       features_vec|       rawPrediction|         probability|prediction|
+--------------------+-----------+-----+--------------------+--------------------+--------------------+----------+
|(2302,[0,1,4,5,64...|          6|  1.0|(2303,[0,1,4,5,64...|[-20.652946698958...|[1.07285051282855...|       1.0|
|(2302,[0,1,4,10,2...|          7|  0.0|(2303,[0,1,4,10,2...|[22.8976498047761...|[0.99999999988632...|       0.0|
|(2302,[0,1,4,11,1...|          6|  0.0|(2303,[0,1,4,11,1...|[19.4843287026586...|[0.99999999654805...|       0.0|
|(2302,[0,1,4,12,1...|          8|  1.0|(2303,[0,1,4,12,1...|[-21.309143076804...|[5.56617360333133...|       1.0|
|(2302,[0,1,4,12,1...|          8|  1.0|(2303,[0,1,4,12,1...|[-17.536798543214...|[2.42027743590876...|       1.0|
|(2302,[0,1,4,12,3...|          5|  1.0|(2303,[0,1,4,12,3...|[-21.575695338535...|[4.26377786893109...|       1.0|
|(2302,[0,1,4,12,3...|          5|  1.0|(2303,[0,1,4,12,3...|[-21.575695338535...|[4.26377786893109...|       1.0|
|(2302,[0,1,4,12,3...|          5|  1.0|(2303,[0,1,4,12,3...|[-21.575695338535...|[4.26377786893109...|       1.0|
|(2302,[0,1,4,12,3...|          5|  1.0|(2303,[0,1,4,12,3...|[-21.575695338535...|[4.26377786893109...|       1.0|
|(2302,[0,1,4,12,3...|          5|  1.0|(2303,[0,1,4,12,3...|[ 21.575695338535...|[4.26377786893109...|       1.0|
|(2302,[0,1,4,12,3...|          5|  1.0|(2303,[0,1,4,12,3...|[-21.575695338535...|[4.26377786893109...|       1.0|
|(2302,[0,1,4,12,3...|          5|  1.0|(2303,[0,1,4,12,3...|[-21.575695338535...|[4.26377786893109...|       1.0|
|(2302,[0,1,4,12,3...|          5|  1.0|(2303,[0,1,4,12,3...|[-21.575695338535...|[4.26377786893109...|       1.0|
|(2302,[0,1,4,12,3...|          5|  1.0|(2303,[0,1,4,12,3...|[-21.575695338535...|[4.26377786893109...|       1.0|
+--------------------+-----------+-----+--------------------+--------------------+--------------------+----------+
```

Exercise   Please evaluate the performance of the logistic regression model using accuracy metrics on test data.

Sequence Embeddings

Let us move on to the second part of this chapter that covers sequence embeddings. We looked at different ways to convert text into numerical form using bag of words, count vectorizer, TF-IDF, etc. However, none of these considers the semantics, context, or order in which the text or neighboring text appears. However, we know for a fact that context and sequence matter a lot for understanding any text data. That's where embeddings can shine and provide a robust mechanism to understand and represent text data in a numerical form, which is more effective compared with other approaches.

Let us take an example to understand sequence embeddings better. We all use mobile phones connected to the Internet all the time. We use so many apps throughout the day like Facebook, Amazon, Twitter, etc. Some of these apps provide relevant content or items to us to keep us engaged, whereas sometimes we struggle finding the right information or product. Similarly, millions of people use the same apps every day, yet each one of them takes a different route/set of steps to seek the relevant information/product. This could be termed as *individual user journey*. Many times, people are left frustrated with the app experience or disappointed due to missing information. In such cases, it becomes difficult to find out if the user was satisfied with the overall experience. The individual user journeys vary from the rest on multiple parameters such as total time spent, total number of pages viewed, liked content/product, or reviewed content/product.

So, if a business were to understand their customers better, they would need to understand which set of user journeys are resulting into less conversion or drop-off vs. which user journeys are more engaging and successful.

Sequence embedding is a powerful way that offers us the flexibility to not only compare any two individual user journeys but also use them to predict the probability of a visitor's conversion.

Embeddings

As mentioned already, the techniques like count vectorizer, TF-IDF, and hashing vectorization do not consider semantic meanings of the text or the context in which words are present. Embeddings are unique in terms of capturing the context of the words and representing it in such a way that words with similar meanings are represented with a similar sort of embeddings. There are two ways to calculate the embeddings:

1. Skip gram

2. Continuous bag of words (CBOW)

Now we would not be going in depth to cover these techniques mentioned, but on a high level, both the methods give the embedding values that are the weights of the hidden layer in a neural network. The embedding vector size can be chosen based on a requirement, but a size of 100 works well for most of the cases. We will make use of word2vec in Spark to create embeddings. We will use a sample retail dataset for this exercise:

```
[In]:from pyspark.sql import SparkSession
[In]:spark=SparkSession.builder.appName('seq_embedding').getOrCreate()
[In]:from pyspark.ml.functions import vector_to_array
[In]:from pyspark.ml.feature import StringIndexer
[In]:from pyspark.sql.window import Window
[In]:import pandas as pd

[In]:file_location = "/FileStore/tables/embedding_dataset.csv"
[In]:file_type = "csv"

[In]:infer_schema = "false"
[In]:first_row_is_header = "true"
[In]:delimiter = ","

[In]:df = spark.read.format(file_type) \
        .option("inferSchema", infer_schema) \
        .option("header", first_row_is_header) \
        .option("sep", delimiter) \
        .load(file_location)

[In]:display(df)
[Out]:
```

	user_id	page	timestamp	visit_number	time_spent	converted
1	8057ed24427be18922f640b20b60997e7d070946b6c8f48117ae4d6dad0ebb23	homepage	2017-05-24T22:00:41.000Z	0	0.16666667	1
2	8057ed24427be18922f640b20b60997e7d070946b6c8f48117ae4d6dad0ebb23	product info	2017-05-24T22:00:51.000Z	0	0.4	1
3	8057ed24427be18922f640b20b60997e7d070946b6c8f48117ae4d6dad0ebb23	product info	2017-05-24T22:01:15.000Z	0	0.31666666	1
4	8057ed24427be18922f640b20b60997e7d070946b6c8f48117ae4d6dad0ebb23	product info	2017-05-24T22:02:42.000Z	0	0.6333333	1
5	8057ed24427be18922f640b20b60997e7d070946b6c8f48117ae4d6dad0ebb23	product info	2017-05-24T22:03:20.000Z	0	0.15	1
6	8057ed24427be18922f640b20b60997e7d070946b6c8f48117ae4d6dad0ebb23	homepage	2017-05-25T21:10:55.000Z	1	0.8333333	1

```
[In]: df.count()
[Out]: 1096955
[In]: df.select('user_id').distinct().count()
[Out]: 104087

[In]: df.printSchema()
[Out]:
```

```
        root
          |-- user_id: string (nullable = true)
          |-- page: string (nullable = true)
          |-- timestamp: string (nullable = true)
          |-- visit_number: string (nullable = true)
          |-- time_spent: string (nullable = true)
          |-- converted: string (nullable = true)
```

As we can see, the dataset contains six columns. It includes the unique user ID, the web page category being viewed, visit number, time spent on the page category, and conversion status. The total number of records in the dataset is close to 1M, and there are 0.1M unique users. All the columns are of the string datatype:

```
[In]: df.groupBy('page').count().orderBy('count',ascending=False).
show(10,False)
[Out]:
```

```
+-------------+------+
|page         |count |
+-------------+------+
|product info |767131|
|homepage     |142456|
|added to cart|67087 |
|others       |39919 |
|offers       |32003 |
|buy          |24916 |
|reviews      |23443 |
+-------------+------+
```

The whole idea of sequence embeddings is to translate the series of steps taken by the user during their online journey into a page sequence that can be used for calculating embedding scores. The first step is to remove any of the consecutive duplicate pages during the journey of a user. We create an additional column that captures the previous page of the user using the window function in PySpark:

```
[In]: w = Window.partitionBy("user_id").orderBy('timestamp')
[In]: df = df.withColumn("previous_page", lag("page", 1, 'started').
over(w))
[In]: df.select('user_id','timestamp','previous_page','page').
show(10,False)
[Out]:
```

```
+------------------------------------------------------------------+------------------------+-------------+------------+
|user_id                                                           |timestamp               |previous_page|page        |
+------------------------------------------------------------------+------------------------+-------------+------------+
|000b57de22d67187b62f4358a063ed49578cee26b49ec99a76b55d77999cb6d1  |2017-10-23T23:02:50.000Z|started      |product info|
|000b57de22d67187b62f4358a063ed49578cee26b49ec99a76b55d77999cb6d1  |2017-10-25T23:00:43.000Z|product info |product info|
|000b57de22d67187b62f4358a063ed49578cee26b49ec99a76b55d77999cb6d1  |2018-01-04T17:58:29.000Z|product info |product info|
|000b57de22d67187b62f4358a063ed49578cee26b49ec99a76b55d77999cb6d1  |2018-01-04T17:58:30.000Z|product info |product info|
|000b57de22d67187b62f4358a063ed49578cee26b49ec99a76b55d77999cb6d1  |2018-01-04T17:58:31.000Z|product info |product info|
|000b57de22d67187b62f4358a063ed49578cee26b49ec99a76b55d77999cb6d1  |2018-01-04T17:58:31.000Z|product info |product info|
|000b57de22d67187b62f4358a063ed49578cee26b49ec99a76b55d77999cb6d1  |2018-01-04T17:58:31.000Z|product info |product info|
|000b57de22d67187b62f4358a063ed49578cee26b49ec99a76b55d77999cb6d1  |2018-01-04T17:58:34.000Z|product info |product info|
|000b57de22d67187b62f4358a063ed49578cee26b49ec99a76b55d77999cb6d1  |2018-02-11T21:37:57.000Z|product info |product info|
|001a13b2d3fae30b92d751c06f1edcfa222b1e488f96f7b9e381fbd423572ceb  |2018-02-19T21:03:20.000Z|started      |homepage    |
+------------------------------------------------------------------+------------------------+-------------+------------+
```

```
[In]:
def indicator(page, prev_page):
    if page == prev_page:
        return 0
    else:
        return 1
```

```
[In]:page_udf = udf(indicator,IntegerType())
[In]: df = df.withColumn("indicator",page_udf(col('page'),col('previous_
page'))) \
        .withColumn('indicator_cummulative',sum(col('indicator')).over(w))
```

Now, we create a function to check if the current page is like the previous page and indicate the same in a new column indicator. indicator_cumulative is the column to track the number of distinct pages during the user journey:

```
[In]: df.select('previous_page','page','indicator','indicator_
cummulative').show(20,False)
[Out]:
```

```
+------------+------------+---------+--------------------+
|previous_page|page        |indicator|indicator_cummulative|
+------------+------------+---------+--------------------+
|started     |product info|1        |1                   |
|product info |product info|0        |1                   |
|product info |product info|0        |1                   |
|product info |product info|0        |1                   |
|product info |product info|0        |1                   |
|product info |product info|0        |1                   |
|product info |product info|0        |1                   |
|product info |product info|0        |1                   |
|product info |product info|0        |1                   |
|started     |homepage     |1        |1                   |
|homepage    |product info|1        |2                   |
|product info |product info|0        |2                   |
|product info |product info|0        |2                   |
|product info |product info|0        |2                   |
|product info |product info|0        |2                   |
|product info |product info|0        |2                   |
|product info |product info|0        |2                   |
```

```
[In]: w2=Window.partitionBy(["user_id",'indicator_cummulative']).
orderBy('timestamp')
[In]:df= df.withColumn('time_spent_cummulative',sum(col('time_spent')).
over(w2))
[In]: df.select('timestamp','previous_page','page','indicator','indicat
or_cummulative','time_spent','time_spent_cummulative').show(20,False)
[Out]:
```

```
+------------------------+------------+-------------+---------+---------------------+-----------+-----------------------+
|timestamp               |previous_page|page        |indicator|indicator_cummulative|time_spent |time_spent_cummulative|
+------------------------+------------+-------------+---------+---------------------+-----------+-----------------------+
|2017-10-23T23:02:50.000Z|started     |product info|1        |1                    |0.08       |0.08                  |
|2017-10-25T23:00:43.000Z|product info|product info|0        |1                    |0.08       |0.16                  |
|2018-01-04T17:58:29.000Z|product info|product info|0        |1                    |0.016666668|0.176666668           |
|2018-01-04T17:58:30.000Z|product info|product info|0        |1                    |0.016666668|0.193333336           |
|2018-01-04T17:58:31.000Z|product info|product info|0        |1                    |0.0        |0.243333336           |
|2018-01-04T17:58:31.000Z|product info|product info|0        |1                    |0.0        |0.243333336           |
|2018-01-04T17:58:31.000Z|product info|product info|0        |1                    |0.05       |0.243333336           |
|2018-01-04T17:58:34.000Z|product info|product info|0        |1                    |0.08       |0.323333336           |
|2018-02-11T21:37:57.000Z|product info|product info|0        |1                    |0.08       |0.40333333600000004   |
|2018-02-19T21:03:20.000Z|started     |homepage    |1        |1                    |0.26666668 |0.26666668            |
|2018-02-19T21:03:36.000Z|homepage    |product info|1        |2                    |0.43333334 |0.43333334            |
|2018-02-19T21:04:02.000Z|product info|product info|0        |2                    |0.083333336|0.516666676           |
|2018-02-19T21:04:47.000Z|product info|product info|0        |2                    |0.15       |0.6666666760000001    |
|2018-02-19T21:05:40.000Z|product info|product info|0        |2                    |0.05       |0.7166666760000001    |
|2018-02-19T21:05:43.000Z|product info|product info|0        |2                    |0.71666664 |1.4333333160000001    |
|2018-02-19T21:06:26.000Z|product info|product info|0        |2                    |0.2        |1.633333316           |
+------------------------+------------+-------------+---------+---------------------+-----------+-----------------------+
```

In the next stage, we calculate the aggregated time spent on similar pages so that only a single record can be kept for representing consecutive pages:

```
[In]: w3 =Window.partitionBy(["user_id",'indicator_cummulative']).
orderBy(col('timestamp').desc())
```

```
[In]: df = df.withColumn('final_page',first('page').over(w3))\
    .withColumn('final_time_spent',first('time_spent_cummulative').
    over(w3))
```

```
[In]: df.select(['time_spent_cummulative','indicator_
cummulative','page','final_page','final_time_spent']).show(10,False)
```

[Out]:

```
+------------------------+---------------------+------------+------------+--------------------+
|time_spent_cummulative  |indicator_cummulative|page        |final_page  |final_time_spent    |
+------------------------+---------------------+------------+------------+--------------------+
|0.40333333600000004     |1                    |product info|product info|0.40333333600000004 |
|0.323333336             |1                    |product info|product info|0.40333333600000004 |
|0.243333336             |1                    |product info|product info|0.40333333600000004 |
|0.243333336             |1                    |product info|product info|0.40333333600000004 |
|0.243333336             |1                    |product info|product info|0.40333333600000004 |
|0.193333336             |1                    |product info|product info|0.40333333600000004 |
|0.176666668             |1                    |product info|product info|0.40333333600000004 |
|0.16                    |1                    |product info|product info|0.40333333600000004 |
|0.08                    |1                    |product info|product info|0.40333333600000004 |
|0.26666668              |1                    |homepage    |homepage    |0.26666668          |
+------------------------+---------------------+------------+------------+--------------------+
```

```
[In]: aggregations = []
[In]: aggregations.append(max(col('final_page')).alias('page_emb'))
[In]: aggregations.append(max(col('final_time_spent')).alias('time_
spent_emb'))
[In]: aggregations.append(max(col('converted')).alias('converted_emb'))

[In]: df_embedding = df.select(['user_id','indicator_cummulative','final_
page','final_time_spent','converted']).groupBy(['user_id','indicator_
cummulative']).agg(*aggregations)

[In]: w4 = Window.partitionBy(["user_id"]).orderBy('indicator_cummulative')
[In]: w5 = Window.partitionBy(["user_id"]).orderBy(col('indicator_
cummulative').desc())
```

Finally, we use collect list to combine all the pages of the user journey into a single list and for time spent as well. As a result, we end with a user journey in the form of a page list and time spent list:

```
[In]:df_embedding = df_embedding.withColumn('journey_page', collect_
list(col('page_emb')).over(w4))\
                        .withColumn('journey_time_temp', collect_
                        list(col('time_spent_emb')).over(w4)) \
                        .withColumn('journey_page_final',first('journey_
                        page').over(w5))\
                        .withColumn('journey_time_final',first('journey_
                        time_temp').over(w5)) \
                        .select(['user_id','journey_page_final','journey_
                        time_final','converted_emb'])
```

Each user is represented by a single journey and time spent vector:

```
[In]: df_embedding = df_embedding.dropDuplicates()

[In]: df_embedding.count()
[Out]: 104087

[In]: df_embedding.select('user_id').distinct().count()
[Out]: 104087
```

```
[In]: df_embedding.select('user_id','journey_page_final','journey_time_
final').show(10)
[Out]:
```

```
+-------------------+-------------------+-------------------+
|            user_id| journey_page_final| journey_time_final|
+-------------------+-------------------+-------------------+
|000b57de22d67187b...|       [product info]|[0.40333333600000...|
|001a13b2d3fae30b9...|[homepage, produc...|[0.26666668, 3.09...|
|0021689e622e8f268...|[homepage, produc...|[0.11666667, 8.26...|
|00323567146f62efb...|[homepage, produc...|[0.28333333, 0.26...|
|003d29d24cff1d994...|[homepage, produc...|[0.25, 0.89666667...|
|00495ae8c90665343...|       [product info]|[1.4733333000000004]|
|004e96d0dc01f2541...|       [product info]|[3.5533333000000002]|
|0058982521702bf10...|       [product info]|        [2.210000036]|
|005d6c72c8c6a6fe1...|[homepage, offers...|[0.48333332, 0.66...|
|005e57f931717c829...|[product info, ot...|[3.8500001, 1.35,...|
+-------------------+-------------------+-------------------+
```

We can now move to create embeddings using the word2vec model by feeding it the user journey sequence. The embedding size is kept to 100 for this part:

```
[In]: from pyspark.ml.feature import Word2Vec
[In]: word2vec = Word2Vec(vectorSize = 100, inputCol = 'journey_page_
final', outputCol = 'embedding')
[In]: model = word2vec.fit(df_embedding)
[In]: result = model.transform(df_embedding)

[In]: result.show(3)
[Out]:
```

```
+-------------------+-------------------+-------------------+-------------+-------------------+
|            user_id| journey_page_final| journey_time_final|converted_emb|          embedding|
+-------------------+-------------------+-------------------+-------------+-------------------+
|000b57de22d67187b...|       [product info]|[0.40333333600000...|            0|[0.04747530445456...|
|001a13b2d3fae30b9...|[homepage, produc...|[0.26666668, 3.09...|            0|[0.00918909907341...|
|0021689e622e8f268...|[homepage, produc...|[0.11666667, 8.26...|            0|[0.00918909907341...|
+-------------------+-------------------+-------------------+-------------+-------------------+
```

We can extract embeddings for each page category using getVectors(), but do ensure to change the datatype of the embeddings to double as the embeddings' original format is vector in Spark:

```
[In]: embeddings=model.getVectors()
[In]: embeddings.printSchema()
[Out]:

            root
             |-- word: string (nullable = true)
             |-- vector: vector (nullable = true)


[In]: embeddings=embeddings.withColumn('vector',vector_to_array('vector'))
[In]: embeddings.printSchema()
[Out]:

   root
     |-- word: string (nullable = true)
     |-- vector: array (nullable = false)
     |      |-- element: double (containsNull = false)


[In]: embeddings.show()
[Out]:

        +-------------+--------------------+
        |         word|              vector|
        +-------------+--------------------+
        | product info|[0.04747530445456...|
        |added to cart|[0.15066237747669...|
        |       others|[0.16220042109489...|
        |          buy|[0.37700653076171...|
        |       offers|[0.01404000166803...|
        |      reviews|[0.22849042713642...|
        |     homepage|[-0.0290971063077...|
        +-------------+--------------------+
```

As we can observe, the vocabulary size is 7 because we were dealing with seven page categories only. Each of these page categories now can be represented with the help of the embedding vector of size 100:

```
[In]: page_categories=embeddings.select('word').distinct().collect()
[In]: unique_pages = [i.word for i in page_categories]
[In]: print(unique_pages)
[Out]:
```

```
['product info', 'added to cart', 'others', 'buy', 'offers', 'reviews', 'homepage']
```

In order to visualize the embeddings of each of the page categories, we can convert the embeddings Dataframe to a Pandas Dataframe and later use matplotlib to plot the embeddings:

```
[In]: pd_df_embedding = embeddings.toPandas()
[In]: pd_df_embedding.head()
[Out]:
```

	word	vector
0	product info	[0.04747530445456505, -0.02869913913309574, 0....
1	added to cart	[0.1506623774766922, -0.17722554504871368, -0....
2	others	[0.1622004210948944, 0.05221569910645485, -0.2...
3	buy	[0.377006530076171875, -0.2556954026222229, -0....
4	offers	[0.014040001668035984, -0.08490971475839615, 0...

```
[In]: X=pd.DataFrame(pd_df_embedding['vector'].values.tolist())
[In]: X.shape
[Out]: (7,100)
[In]: X.head(3)
[Out]:
```

	0	1	2	3	4	5	6	7	8	9 ...	90	91	92	93	94	95	96	97	98	99
0	0.047475	-0.028699	0.065521	0.014660	0.013824	0.181943	-0.027661	0.037527	0.123778	-0.049622 ...	0.079849	-0.115598	-0.237508	-0.005171	-0.165508	-0.106722	0.033441	-0.000553	0.099693	0.090593
1	0.150662	-0.177226	-0.148702	0.142573	0.026625	0.096106	0.025328	-0.105925	0.109929	0.186005 ...	-0.044250	0.002983	-0.223052	0.122902	-0.062829	0.083307	0.141899	0.107690	-0.084157	0.067637
2	0.162200	0.052216	-0.209934	-0.054830	-0.044091	0.194070	-0.055435	-0.086015	-0.141650	0.054387 ...	0.165940	0.159493	0.129481	0.085579	0.201789	0.132260	0.141637	0.063040	0.065906	0.065104

In order to better understand the relation between these page categories, we can use the dimensionality reduction technique (PCA) and plot these seven page embeddings on a two-dimensional space:

```
[In]: from sklearn.decomposition import PCA
[In]: pca = PCA(n_components=2)
[In]: pca_df = pca.fit_transform(X)
[In]: pca_df
[Out]:

                 array([[-0.61822842,  0.16885239],
                        [ 0.39320573, -0.46453367],
                        [-0.0721883 ,  0.15486883],
                        [ 1.10692327, -0.9131682 ],
                        [-0.89615923, -0.48067825],
                        [ 0.69542943,  1.38978311],
                        [-0.60898248,  0.1448758 ]])


[In]: import matplotlib.pyplot as plt
[In]: %matplotlib inline
[In]: plt.figure(figsize=(10,10))
[In]: plt.scatter(pca_df[:, 0], pca_df[:, 1])

[In]: for i,unique_page in enumerate(unique_pages):
        plt.annotate(unique_page,horizontalalignment='right',vertical
        alignment='top',xy=(pca_df[i, 0], pca_df[i, 1]))

[In]: plt.show()
```

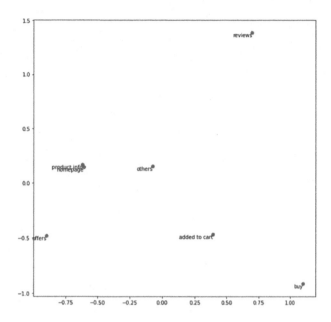

As we can clearly observe, the embeddings of product info and homepage are near to each other in terms of similarity. Offers and reviews are very far when it comes to representation through embeddings. These individual embeddings can be combined and used for user journey comparison and classification using Machine Learning.

Conclusion

In this chapter, we covered the steps to do text processing using PySpark and creating sequence embeddings for representing online user journey data.

Overall in the book, we looked at multiple sets of algorithms to solve different problems right from linear regression to building a recommender system in PySpark. Again, we used some standard datasets (small and mid-size), but the same codebase could be applied on big datasets as well without changing too many things. Spark provides the flexibility to create customized pipelines as per requirement of the workflow. Now sometimes Spark might be an overkill to solve a small problem or for a small POC (proof of concept) because Pandas and sklearn could handle that much data. Hence, one should carefully evaluate the resource and tool landscape before starting to write code. There could be a bunch of metrics on which the right framework could be chosen such as size of data, scale of the project, timelines, latency rate, resources available, costs, etc. The bottom line is that one can leverage the power of Spark to deal with large data and build scalable ML models very quickly.

Index

A, B

ALS method, 187
Alternating least squares (ALS), 172

C

Clustering
  agglomerative, 142–144, 146
  approaches, 128
  centroids, 134–140
  correlation coefficient, 152, 153
  databricks notebooks, 147–150
  definition, 127
  elbow method, 141
  Euclidean method, 129
  hierarchica, 142
  intra-cluster distance, 151
  K-means, 129–131, 133, 134
  3D visualization, 155
ClusteringEvaluator method, 151
Code, LR
  Dataframe, 70
  dataset, 88, 90–92, 94, 96–98
  Jupyter notebook, 66
  output variable, 69
  Pyspark, 72
  RMSE, 71, 73
  statistical measures, 67, 68
  VectorAssembler, 69
Collaborative filtering-based RS
  decisions, 162
  explicit feedback, 164

  implicit feedback, 164
  latent factor, 170–173
  missing values, 166, 168, 170
  nearest neighbors, 165
  user item matrix, 163, 164
"columns" method, 19
Confusion matrix, 84
  accuracy, 100
  precision, 101
  recall, 100
Content-based RS
  cosine similarity, 162
  Euclidean distance, 161
  Movie attributes, 160
  user profile, 161
Continuous bag of words (CBOW), 204
Corpus, 190
corr function, 69
CountVectorizer method, 194

D

Databricks notebook, 14
dropDuplicates function, 32

E, F

Elbow method, 141
Euclidean method, 128

G

Graph computation, 8
groupBy function, 118, 121, 153, 202

© Pramod Singh 2022
P. Singh, *Machine Learning with PySpark*, https://doi.org/10.1007/978-1-4842-7777-5

Printed in the United States
by Baker & Taylor Publisher Services